SYSTEMATIC BOTANY MONOGRAPHS

VOLUME 62

Monograph of Aciotis (Melastomataceae)

Alina Freire-Fierro

THE AMERICAN SOCIETY OF PLANT TAXONOMISTS
15 July 2002

SYSTEMATIC BOTANY MONOGRAPHS

ISSN 0737-8211

ISBN 0-912861-62-2

Printed in the United States of America

Publication of Volume 62 of Systematic Botany Monographs was partly funded by Walter L. Meagher, whose generous contribution is gratefully acknowledged.

MONOGRAPH OF ACIOTIS (MELASTOMATACEAE)

Alina Freire-Fierro
Missouri Botanical Garden
P. O. Box 299
St. Louis, Missouri 63166-0299

ABSTRACT. *Aciotis* (Melastomataceae) is a genus of 13 species of herbs and subshrubs that occur from southern Mexico to northern Bolivia and southeastern Brazil. Traditionally, *Aciotis* has been placed in Tibouchineae [=Melastomeae s.l.], because it has capsular fruits; yet, some species of *Aciotis* have also baccate fruits. Most of the Melastomeae have tuberculate seeds, but all species of *Aciotis* have foveolate seeds. Three species (*A. rubricaulis*, *A. indecora*, and *A. purpurascens*) have thyrsoid inflorescences and indehiscent baccate fruits; four species (*A. annua*, *A. ornata*, *A. polystachya*, and *A. wurdackiana*) have cymose inflorescences and dry dehiscent capsules. The remaining six species (*A. acuminifolia*, *A. circaeifolia*, *A. ferreirana*, *A. olivieriana*, *A. paludosa*, and *A. viscida*) exhibit several combinations of characters, such as thyrsoid inflorescences with capsular fruits or cymose inflorescences with indehiscent baccate fruits. Illustrations and distribution maps are provided for all species.

INTRODUCTION

Aciotis D. Don (Melastomataceae) is a neotropical genus of 13 species. It includes herbs and subshrubs with actinomorphic tetramerous flowers less than 1.5 cm long. The eight stamens lack the dorsal connective projections usually present in other genera of Melastomataceae, and the ovary is mostly bilocular (very rarely trilocular). The fruit is capsular in several species, but fleshy and indehiscent in others; the seeds are cochleate with a foveolate testa. Fruit and inflorescence features seem to divide the genus into two groups. One group includes species with paniculate to thyrsoid inflorescences and baccate fruits, and the other group includes species with cymes to double cymes and capsular fruits; however, some species bridge this apparent division and cannot be placed in either group.

Aciotis is considered a natural genus (Almeda 1997a). Although it is readily separated from other genera, recognition of species boundaries has always been a problem. Wurdack (1980) said of his treatment of the Ecuadorian species of *Aciotis* that it "is as makeshift as that previously published for Venezuela [Wurdack 1973], [and] species distinctions [are] often fuzzy." After a detailed analysis of morphological characters obtained from herbarium material and from field observations, it was possible to identify more reliable and diagnostic characters for distinguishing species, and the present monograph provides descriptions and a key using characters that are, for the most part, easy to discern.

Aciotis belongs to the tribe Melastomeae Bartl., which now (Renner 1993; Clausing & Renner 2001) also includes genera previously included in Tibouchineae Cogn. (Triana 1873; Cogniaux 1891). The placement of *Aciotis* in Melastomeae is based on the presence of cochleate seeds, pedoconnectives on the stamens (i.e., extension of the connectives below the anthers) (Fig. 5), and trichomes on the ovary apex; however, some species of *Aciotis* lack pedoconnectives and others trichomes on the ovary apex. Although

Melastomeae is also defined by the capsular fruits and tibouchinoid seeds (e.g., cochleate and tuberculate), some species of *Aciotis* have fleshy fruits, and the entire genus, as well as species of other genera of Melastomeae, such as *Macairea* DC., *Ernestia* DC., *Comolia* DC., *Acisanthera* P. Br., *Microlepis* (DC.) Miq,. and *Svitramia* Cham., have foveolate seeds (Whiffin & Tomb 1972).

Aciotis is restricted to tropical America. It ranges from southern Mexico and the Lesser Antilles to central Bolivia and southeastern Brazil; the eastern lowlands of the Andes and the Guianas are particularly rich in species. The species commonly occur in disturbed habitats, such as abandoned fields and secondary forests. They also grow in wetlands, creeks, and swamps, and less often in primary forests or in dry sandy areas.

Most of the species of *Aciotis* with fleshy and baccate fruits have a wider distribution than those with capsules. Of the species with fleshy indehiscent fruits, three (*A. rubricaulis*, *A. indecora*, and *A. purpurascens*) range from Central America to Bolivia and southeastern Brazil, whereas *A. circaeifolia* and *A. ferreirana* are restricted to Amazonia, and *A. paludosa* to eastern/southeastern Brazil. The widespread *A. acuminifolia*, which has fleshy "capsules," is found in South America and the Lesser Antilles. The species with dry, capsular fruits, *A. olivieriana*, *A. ornata*, *A. polystachya*, *A. viscida*, *A. wurdackiana*, and *A. annua* are only distributed in South America.

During the study several collections were noted to show characteristics intermediate between species. These variants may represent the results of hybridization.

MATERIALS AND METHODS

Morphology. This revision is based on the study of herbarium material complemented by collections made in the field by the author. Approximately 4000 specimens from 50 herbaria representing more than 3000 different collections were examined (see Acknowledgments). For the description of vegetative and reproductive structures, I have used standard and specialized terminology (Sell & Cremers 1987; Weberling 1988; Stearn 1992). Information obtained from herbarium specimens has been entered into the computer program Hypertaxonomy (Skov 1989, 1990), which is designed for monographic or floristic works, including distribution maps and collections lists.

Images of types from U were obtained from the internet [Nationaal Herbarium Nederland (NHN) 1999]. Synonymies for the majority of names were based on personal studies of the types, but some were taken from the literature (Wurdack 1980, Wurdack et al. 1993).

Cytology. Young floral buds were collected in the field, fixed in modified Carnoy's solution (Almeda & Chuang 1992), and then refrigerated. From these buds I obtained the haploid chromosome number from pollen mother cells (PMC). Anthers were placed on a slide, immersed in a droplet of aceto-carmine at 10%; those to be studied were teased apart to release their pollen mother cells. The slide was then warmed and observed at 100× magnification. In instances when the cells were overstained, the slide was destained with 45% acetic acid in ethanol. Once the chromosomes were visible, they were drawn using a camera lucida at 2100× magnification.

TAXONOMIC HISTORY

The first account of species now included in *Aciotis* was by Ruiz and Pavón, who described *Arthrostema caulialatum* [=*Aciotis rubricaulis*] for their *Flora peruviana, et chilensis* (1798–1802), but in volume 4, which remained unpublished until 1954–1957. Copies of the illustrations were circulated by 1830 (Stafleu & Cowan 1983; Greuter 2001), and the plate depicting *Arthrostema caulialatum* includes sufficient detail that the name was thereby effectively published. Aublet (1775) described *Melastoma purpurascens* [=*Aciotis purpurascens*], based on one of his baccate-fruited collections from French Guiana. Bonpland (1818, 1821) later named several species now considered to belong to *Aciotis* in *Rhexia* L. Soon after, David Don (1823) proposed the genus *Aciotis*, defined, among other morphological characteristics, by "baccae globosae, purpurei, grati acidi edules," and the species *Aciotis discolor* [=*Aciotis purpurascens*], based on a collection made by Alexander Anderson, apparently in St. Vincent.

In his *Prodromus*, A. P. de Candolle (1828) maintained Don's monotypic genus *Aciotis* and assigned elsewhere species now included in *Aciotis*. He defined *Rhexia* as having tubular calyces, 4-loculicidal capsules, and confined to temperate North America; he placed species with globose calyces, 2–3-loculicidal capsules, and distributed in the Neotropics in the new genus *Spennera*, first proposed in manuscript by Martius. Some species, now known to be baccate, were placed in *Spennera*, because they had been described erroneously as capsular (e.g., *Aciotis circaeifolia*) or were unknown in fruit (e.g., *Aciotis indecora*). Candolle moved Aublet's baccate *Melastoma purpurascens* to *Miconia* Ruiz & Pav.

Subsequent authors (Bentham 1840; Miquel 1840, 1844, 1851; Naudin 1850) described additional species in *Spennera*. Triana (1871) redefined *Aciotis*, considering it a capsular-fruited genus, and, following primarily the treatments of Candolle and Naudin, he transferred almost all of the species of *Spennera* to *Aciotis*. Yet, despite of his circumscription of *Aciotis*, he included the baccate-fruited *Melastoma purpurascens* in *Aciotis*. Cogniaux (1885) and subsequent authors have followed Triana in defining *Aciotis* as a capsular-fruited genus.

Krasser (1893) in his treatment of the family for *Die Pflanzenfamilien* recognized two sections in *Aciotis*: sect. *Euaciotis* [=*Aciotis*] and sect. *Spennera*, based on anther shape and length of the pedoconnective.

In the most recent study of *Aciotis* prior to my own, Mathies (1981), accepted thirty species and two varieties, and proposed eight new species. Of the new taxa, only two have been validated (Freire-Fierro 2001b). Almeda (1997a) published cytological data and nomenclatural notes for the Mesoamerican species.

The genus has been treated in several floristic works, e.g., those of Panama (Gleason 1958), Guatemala (Standley & Williams 1963), Costa Rica (Standley 1938), Nicaragua (Almeda 2001), Venezuela (Wurdack 1973), the Venezuelan Guayana (Freire-Fierro 2001a), the Guianas (Wurdack et al. 1993), Guyana (Gleason 1932), Suriname (Gleason 1935; Görts-van Rijn & Jansen-Jacobs 1988), French Guiana (Lemée 1953), Santa Catarina, Brazil (Wurdack 1962), Ecuador (Wurdack 1980), and Peru (Macbride 1941).

MORPHOLOGY

Vesture. In most species the vegetative parts are covered with multicellular, glandular or eglandular trichomes, but *A. purpurascens* and many specimens of *A. acuminifolia* are glabrous. It seems that in the latter species pubescence density is associated with habitat. Plants growing in flooded and swampy areas are glabrous, whereas plants in drier areas are glandular or eglandular-pubescent. *Aciotis paludosa* and some representatives of *A. acuminifolia* have abundant, reddish, and long (ca. 1 mm long) eglandular trichomes on the petioles and young stems, and representatives of *A. polystachya*, *A. olivieriana*, *A. indecora*, and *A. viscida* are covered by glandular trichomes. *Aciotis circaeifolia* also has a few scattered glandular hairs, but these are much longer (>1.5 mm long) than those of the previous four species.

Sclereids. In addition to trichomes, *A. annua*, *A. ornata*, *A. polystachya*, and *A. wurdackiana* show web-like striations on the lower side of the leaf that give it a whitish appearance. This feature is possibly due to filiform sclereids, which are present throughout the mesophyll and in the parenchyma of other structures, such as petioles, stems, and pericarps. Solereder (1908) referred to these structures as "spiral tracheids" and observed them in *A. annua*. According to Metcalfe and Chalk (1972), these "spiral tracheids" on the mesophyll occur also in the tribes Sonerileae Triana (*Sonerila* Roxb.), and Miconieae (*Bellucia* Neck. ex Raf. and *Henriettea* DC). Although these cells have a spiral appearance once they have been stretched during the dissection, they do not seem to be part of the vascular tissue; they are usually located in leaf areas where there are no veins. These structures, because of their shape and distribution in the mesophyll extending beneath the epidermal layers (see Figs. 22–23 in Rao & Bhupal, 1973), are most likely filiform sclereids and have been recorded for species of *Mouriri* Aubl. (Morley 1976) and *Memecylon* L. (Fig. 1) (van Thieghem 1891; Solereder 1908; Rao & Jacques-Félix 1978). These two genera have been often included in Melastomataceae. Although Renner (1993) accepts Candolle's (1828) circumscription and reinstated them in their own family Memecylaceae DC., Clausing and Renner (2001) state that "the question whether Memecylaceae should be included in Melastomataceae or ranked as a family, reduces to a matter of ranking and pragmatics of family identification."

Habit. All species of *Aciotis* are herbaceous, and range from 2 cm tall in *A. acuminifolia* to up to 2 m tall in *A. purpurascens* and *A. indecora*. Some species may be annuals, whereas others, especially those with stolons (e.g., *A. ornata*, *A. purpurascens*, and *A. rubricaulis*), seem to be perennials.

Stems. The stems, as in many species of Melastomataceae, are quadrangular. All species bear wings along the edges of the quadrangular stems, and although some species have pronounced wings (e.g., *Aciotis annua*, *A. rubricaulis,* and *A. purpurascens*), others have inconspicuous wings (e.g., *A. ferreirana*, *A. circaeifolia*, and *A. viscida*).

Leaves. The opposite and isophyllous leaves of *Aciotis* are mostly ovate, but they are elliptic to narrowly elliptic in *A. rubricaulis*, *A. purpurascens*, and *A. wurdackiana*. Most species have leaves with a cordate to truncate base, but nearly all representatives of *A. acuminifolia* have leaves with the base typically attenuate to decurrent. The apex is mostly acute to acuminate, but in *A. ornata* and *A. paludosa* the leaf apex typically is obtuse. The leaf blades can be membranous or papyraceous. Although this last character is difficult to detect in herbarium material, it is evident in the field. When membranous leaves have dried, they become translucent; papyraceous leaves, however, appear darker and thicker on the herbarium sheets. The number of primary veins in *Aciotis*, as in most species of

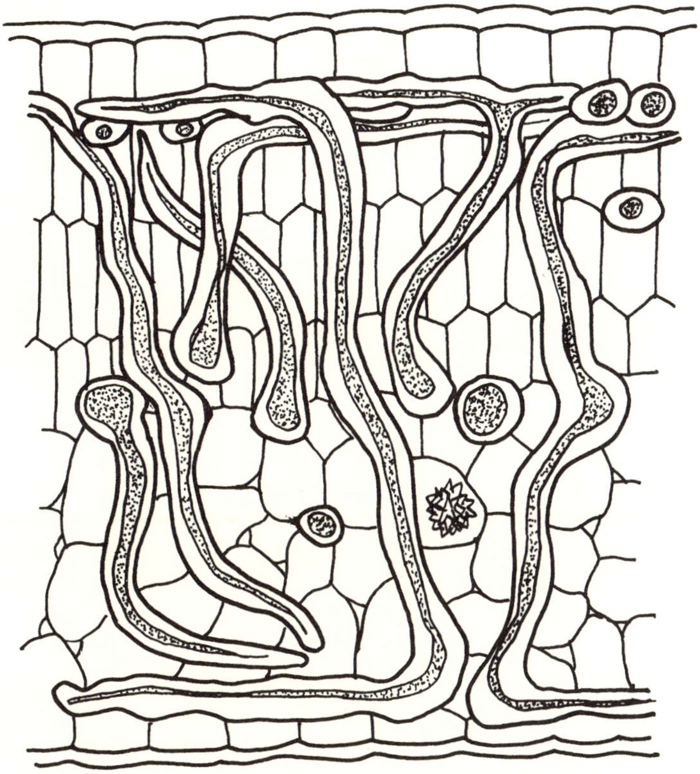

FIG. 1. Transversal section of a leaf of *Memecylon ramiflorum*, showing the filiform sclereids that cross the parenchyma. (After van Tieghem, 1891)

Melastomataceae, is generally five to seven, although most leaves of *A. acuminifolia* and *A. paludosa* have three primary veins. The margin of the leaves is commonly serrulate, but in *A. ferreirana* it is crenate-serrate and in some specimens of *A. annua* doubly serrulate. Leaf damage caused by herbivores is evident in most of the species; however, *A. annua*, *A. ornata*, *A. polystachya*, and *A. wurdackiana* do not show high levels of injury, perhaps because the filiform sclereids distributed in the mesophyll are a deterrent.

Inflorescences. Flowers in *Aciotis* are organized in several kinds of determinate inflorescences (Fig. 2), and were recorded for the French Guianan species of *Aciotis* by Sell and Cremers (1987). Because there are several degrees of arrangements on inflorescences in *Aciotis*, I have followed the terminology that Weberling (1988) used for all the families in Myrtales. The abundance of flowers on the lateral paraclades of the inflorescence differs among species, and although difficult to discern in some specimens, as a general rule it is possible to distinguish two patterns. The first group of species (*Aciotis acuminifolia*, *A. olivieriana*, *A. ornata*, *A. paludosa*, and A. *polystachya*) has lax inflorescences with just one to three pairs of lateral paraclades of the first order and long hypopodia (Fig. 2; see also Sell & Cremers, 1987, Fig. 3b); the outline of the inflorescence is typically rectangular. The second group of species (*A. rubricaulis*, *A. circaeifolia*, *A. ferreirana*, *A. indecora*, *A. purpurascens*, and *A. viscida*) has elongated inflorescences with three to seven pairs of paraclades of the first order and short hypopodia. In this case, the outline of the inflorescence is conical (Fig. 2; see also Sell & Cremers, 1987, Fig. 6a, b). The first group usually has cymes or double biparous cymes, and the second group has thyrses or panicles. Bracts in *Aciotis* are either evident and large (>2 mm long) (e.g., *A. annua* and *A. wurdackiana*), or barely visible and short (0.4–1.2 mm long), as observed in *A. ornata*.

Flowers. *Aciotis* has small (<1.5 cm long) perigyneous actinomorphic flowers (Fig. 3). Before and after anthesis, the flowers seem closed because of the ascendant position of the petals. During anthesis the petals take a perpendicular position in relation to the hypanthium. It is very difficult to press these opened flowers, because the petals are very delicate and fall off very easily.

Calyx. The four sepals are apically free and basally connate; they are abaxially glabrous or covered by glandular or eglandular trichomes. The base of the sepals is totally adnate to the hypanthium, which is apically free and basally adnate to the ovary. The abaxial surface of the hypanthium is glabrous (*A. purpurascens*), or eglandular to glandular-pubescent (e.g., *A. rubricaulis*, *A. acuminifolia*), or entirely covered by glandular trichomes (e.g., *A. polystachya*. *A. indecora,* and *A. viscida*). The adaxial surface of the free portion is usually glabrous (Fig. 4), or it can be covered by glandular trichomes in most of the representatives of *A. wurdackiana*, *A. polystachya*, and *A. olivieriana*. The hypanthium of *A. acuminifolia*, *A. rubricaulis*, A. *purpurascens*, and *A. indecora* is thick-walled and fleshy in fresh mature fruits, and it is easily detachable from the base of the ovary; however, hypanthia become membranous and translucent after they dry out. In these four species, and in *A. circaeifolia*, *A. ferreirana,* and *A. paludosa*, the dried hypanthium shows the size and position of the seeds inside. The hypanthium is coriaceous or papyraceous in most of the species with capsular fruits. The hypanthia of *A. polystachya*, *A. annua*, *A. wurdackiana*, and *A. ornata* have filiform sclereids in the mesophyll; the hypanthia of *A. olivieriana* and *A. viscida* lack these structures. Most of the species have terete hypanthia, but *A. acuminifolia* and *A. olivieriana* have angled hypanthia with eight evident longitudinal costae.

Corolla. (Fig. 3). The four petals are free, alternate with the calyx lobes, are inserted on the apical portion and adaxial side of the hypanthium, and can be white (e.g., *Aciotis rubricaulis* and *A. purpurascens*) or pinkish with a darker longitudinal central line (e.g., *A. polystachya*). The petals are usually shorter than 4 mm, but in *A. circaeifolia* and some individuals of *A. rubricaulis* the petals can be up to 8 mm long. The surface of the petals is mostly glabrous, except for a terminal glandular trichome, but *A. indecora*, *A. polystachya*, and *A. viscida* have petals with the abaxial surface covered by glandular trichomes or sessile glands. Although petal shape seems to be a good character for distinguishing species, it has not been used; petals are very delicate once they are dried, and therefore it is very difficult to dissect them completely from rehydrated flowers. Still, even without dissecting the flowers, it is possible to observe the presence or absence of trichomes on the abaxial surface of the petals in herbarium material.

Androecium (Figs. 4, 5). The eight stamens are disposed in two cycles and lack conspicuous dorsal or ventral appendages; however, a pedoconnective (Fig. 5) is present in some species, although it is not geniculate. Usually, when the pedoconnective is less developed, the anthers are longer than wide and elliptic, but when it is well developed, the

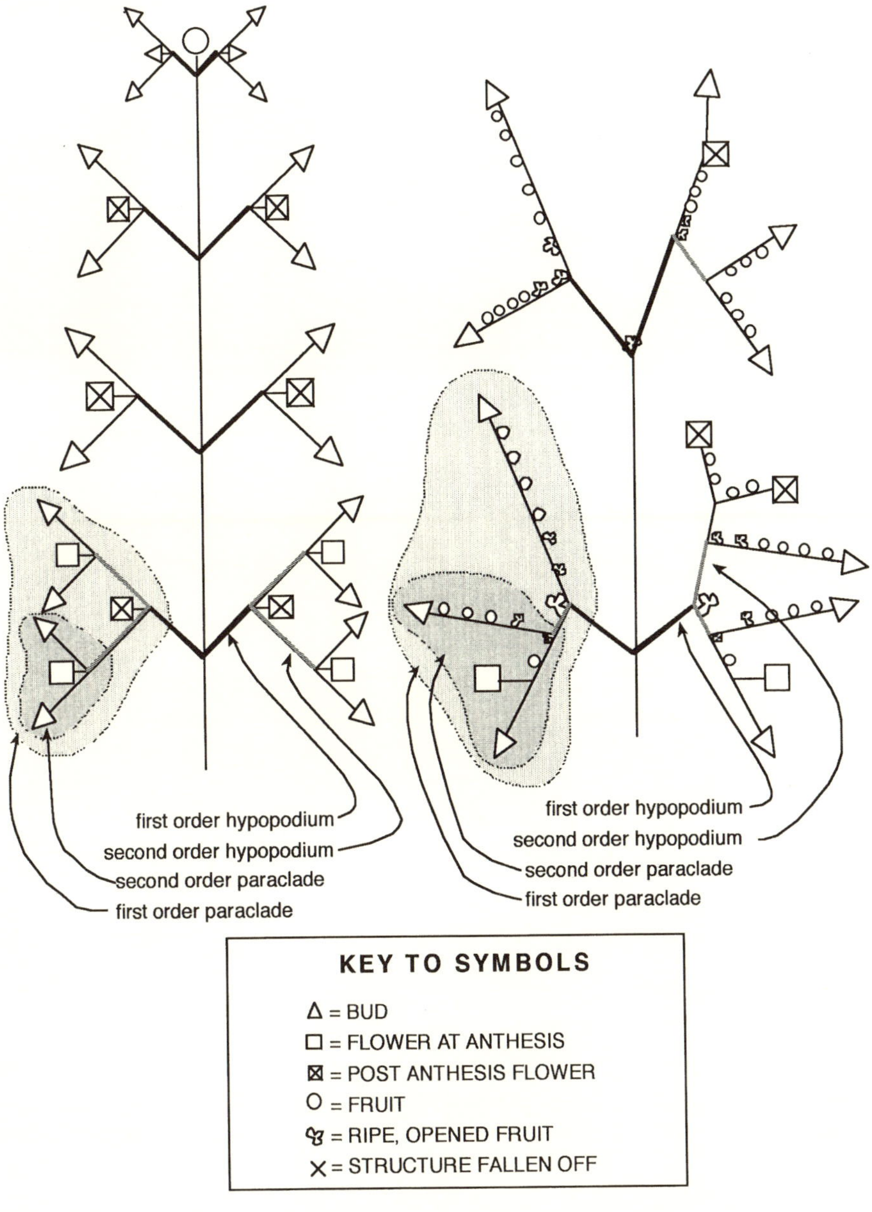

FIG. 2. Inflorescence types in *Aciotis*. Left: inflorescence with four pairs of paraclades of the first order; right: inflorescence with two pairs of paraclades of the first order.

anthers tend to be as long as wide and ovoid. The size of the pedoconnectives (0.1 to 0.6 mm long) and the shape of the anthers (elliptic to ovoid) have been previously used as characters for distinguishing groups of species (Krasser 1893; Mathies 1981; Wurdack et al. 1993); however, because there is continuity from one state to another, there is no way of accurately contrasting these states, and thus I have not used these characters in my key. Anther color, on the other hand, seems to be an informative character; some species have

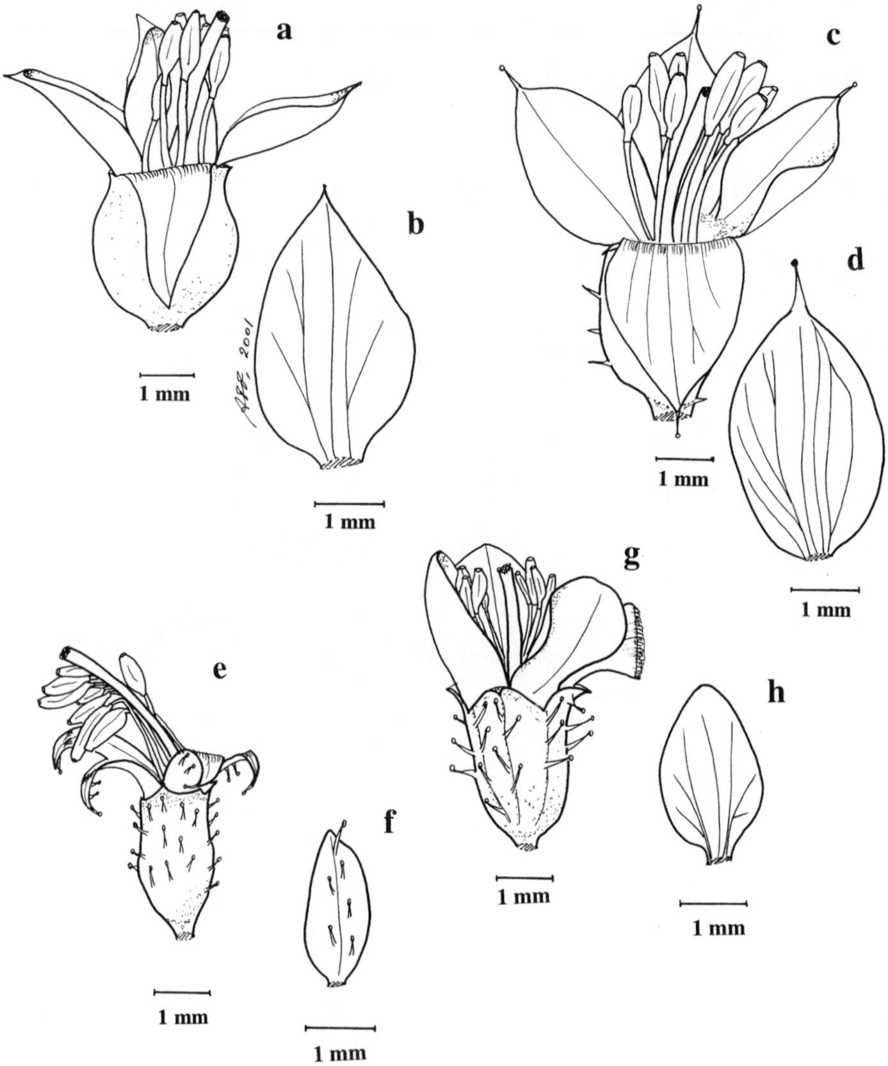

FIG. 3. Flowers and petals of *Aciotis*. a, b. *A. purpurascens*. c, d. *A. rubricaulis*. e, f. *A. polystachya*. g, h. *A. acuminifolia*.

white to light pink anthers (e.g., *Aciotis rubricaulis*, *A. purpurascens*) and others purple anthers (e.g., *A. polystachya*). Although it is a good field character, anther color is sometimes not evident in herbarium material and usually is not noted on the specimen labels.

Gynoecium. The ovary is half-inferior, and it is primarily bilocular with a terminal style and with basal-axile placentation. The ovary wall is adnate to the hypanthium on its inferior half, becoming undifferentiated from the hypanthium, especially when dry. The ovary wall-hypanthium are chartaceous with filiform sclereids in *Aciotis annua*, *A. ornata*, *A. polystachya,* and *A. wurdackiana*; coriaceous in *A. olivieriana* and *A. viscida*; and

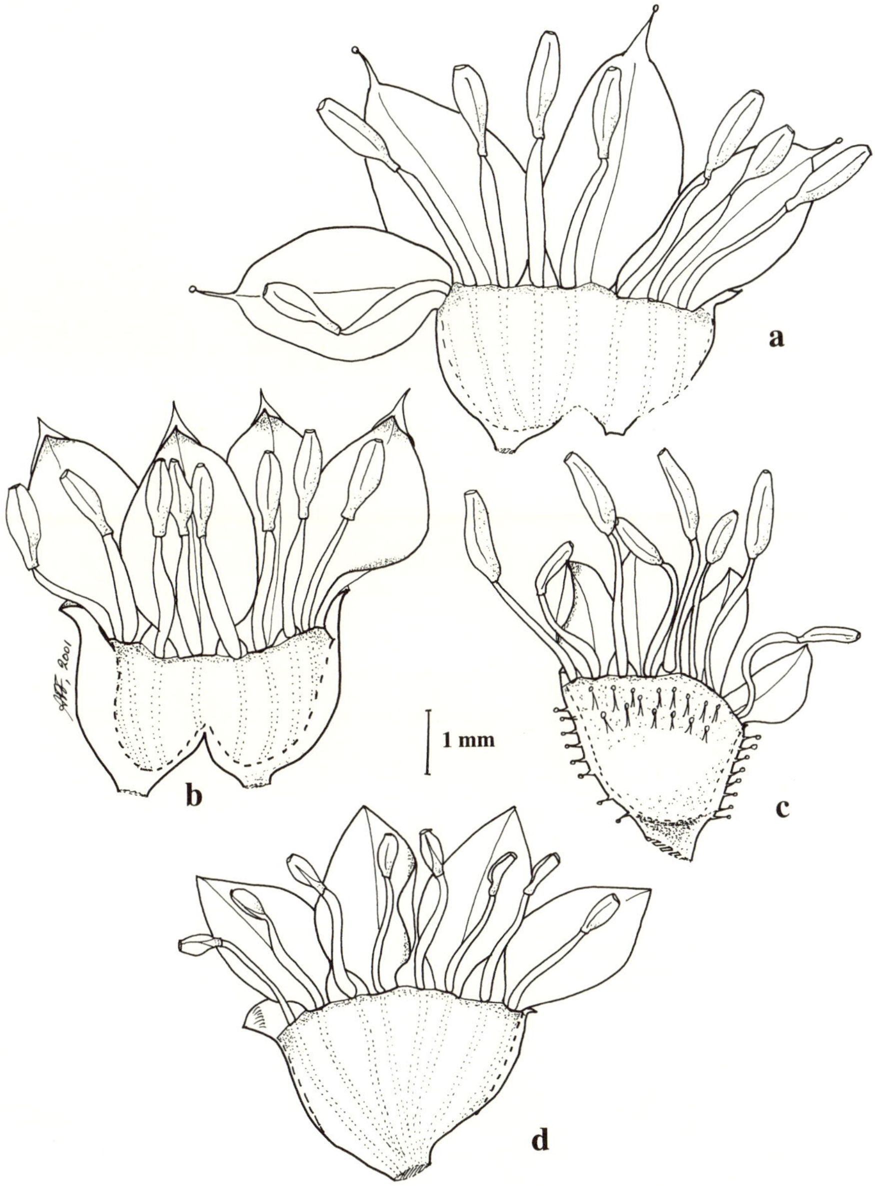

FIG. 4. Flowers of *Aciotis*, showing the internal surface of longitudinally sectioned hypanthia (ovaries and styles removed). a. *A. purpurascens*. b. *A. rubricaulis*. c. *A. polystachya*. d. *A. acuminifolia*.

fleshy (membranous and translucent in herbarium material) in *A. acuminifolia*, *A. rubricaulis*, *A. circaeifolia*, *A. ferreirana*, *A. indecora*, *A. paludosa*, and *A. purpurascens*. The ovary apex is glabrous or beset with glandular trichomes (in *A. annua*, *A. wurdackiana*, *A. ornata*, *A. polystachya*, *A. olivieriana*, *A. indecora,* and *A. viscida*).

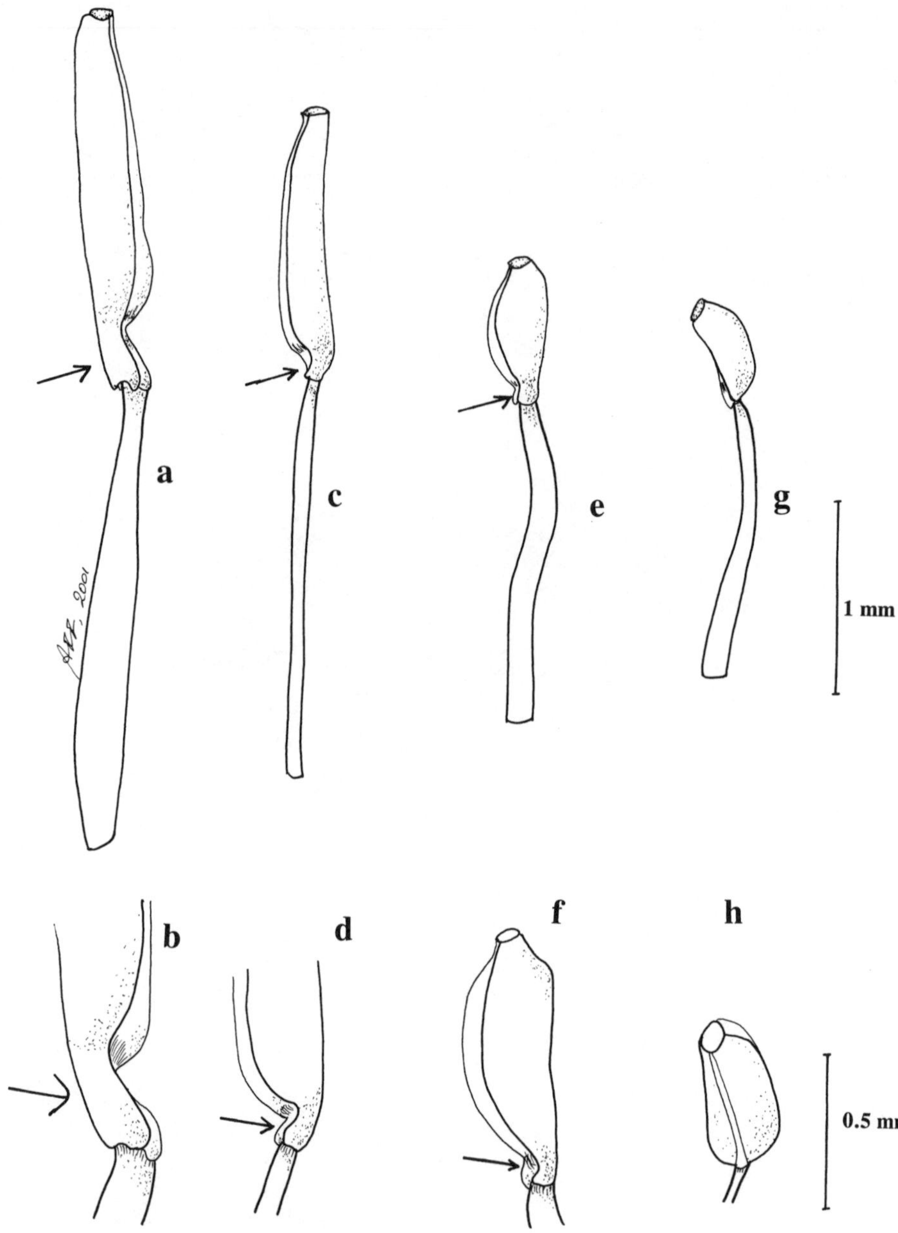

FIG. 5. Stamens of *Aciotis* (arrow points to the pedoconnective). a, b. *A. purpurascens*. c, d. *A. polystachya*. e–h. *A. acuminifolia*.

Fruit. Most treatments of Melastomataceae describe the fruits of *Aciotis* only as capsular (but see Renner, 1993); however, the genus presents three kinds of fruits: dry capsules, fleshy "capsules," and fleshy indehiscent or baccate fruits (Fig. 6). Six species (*Aciotis annua*, *A. olivieriana*, *A. ornata*, *A. polystachya*, *A. viscida*, and *A. wurdackiana*) have septicidal, dehiscent, dry capsules with persistent placentas and pericarps. One species,

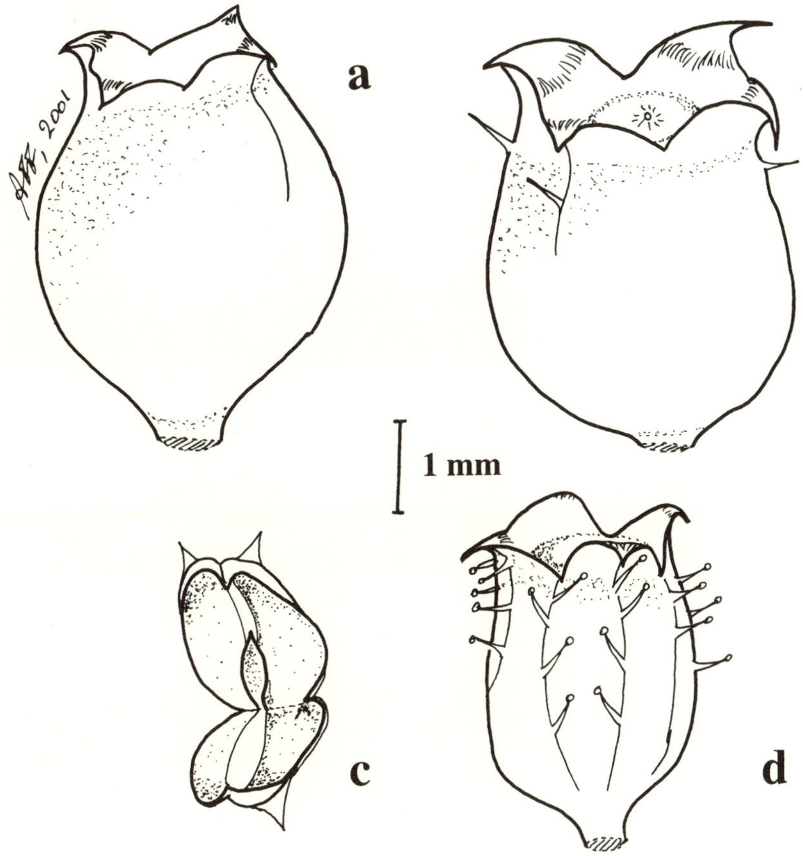

FIG. 6. Fruits of *Aciotis*. a. *A. purpurascens*. b. *A. rubricaulis*. c. *A. polystachya*. d. *A. acuminifolia*.

A. acuminifolia, has also septicidal, dehiscent, fleshy "capsules" with persistent placentas but caducous pericarps after dehiscence. It is easy to observe the persistent placenta in herbarium material and in the field. This character, previously ignored, proved to be of a great taxonomic value. The remaining six species (*A. rubricaulis*, *A. circaeifolia*, *A. ferreirana*, *A. indecora*, *A. paludosa*, and *A. purpurascens*) have baccate fruits. Baumgratz (1983–1985) defines as "bacáceos" (here translated as "baccate) those fruits with fleshy and thin pericarps, fleshy placentas and septa, and seeds not submerged in any kind of pulp. Because these morphological characteristics are also found in the six species of *Aciotis* just listed, this term has been adopted for their fruits. These fruits are fleshy when fresh, but they appear translucent when dried. Transitions between dry and fleshy fruits within the same genus have also been observed in the Old World genera *Melastoma* L. and *Pternandra* W. Jack (Don 1823; Renner 1993; Clausing et al. 2000). Because phylogenetic studies in the family (Renner 1993; Clausing et al. 2000) consider dry capsular fruits as the plesiomorphic state, the baccate fruits of some of the species of *Aciotis* would therefore be considered apomorphic.

All species of *Aciotis* have bilocular fruits, but *A. polystachya* may have bi- or trilocular fruits even on the same individual [e.g., *Freire-Fierro & León 2744* (MO)]. The baccate fruits of *A. purpurascens* and *A. rubricaulis* are buoyant and have spongy parenchyma (pers. obs.), which probably aids in water dispersal.

Seeds. The size of the seeds varies with the kind of fruits. Capsular fruits have small seeds, 0.20–0.50 × 0.15–0.40 mm, while bigger seeds are found in the baccate fruits, 0.45–0.70 × 0.30–0.50 mm. The seeds are cochleate, a character used by Don (1823) and subsequent students of the family to place *Aciotis* in the tribe Melastomeae s.l. (incl. Tibouchinoideae). Although many of the cochleate seeds of Melastomeae are tuberculate, the seeds of *Aciotis* are foveolate. A morphological study of capsular-fruited melastomes (Whiffin & Tomb 1972) describes the seeds of *Aciotis* and other members of Neotropical Melastomeae as "microlicioid" (i.e., short to cochleate and foveolate) and not "tibouchinoid" (i.e., cochleate to elongate-cochleate and tuberculate). For their study, Whiffin and Tomb included specimens of *A. acuminifolia*, *A. annua*, *A. circaeifolia*, *A. indecora*, and *A. rubricaulis*. Central American representatives of the genus have also cochleate and foveolate seeds (Almeda 1997a). The fresh seeds of *A. purpurascens* and *A. rubricaulis* have a bladder-like external surface (pers. obs.), which probably aids in water dispersal.

CHROMOSOME NUMBERS

The most common chromosome numbers in Melastomeae are thought to be n= 9, 10, 11, 12, and dysploid or polyploid derivatives (Almeda & Chuang 1992). The probable base chromosome number for *Aciotis* appears to be x = 10 (Almeda 1997a), and counts of n = 10, 20, 26, 30, and 40 have been reported (see Appendix 1). There are five patterns of chromosome evolution in Melastomeae (Almeda & Chuang 1992; Almeda 1997b): chromosome constancy at the diploid level (*Brachyotum*); intrageneric polyploidy (*Tibouchina* Aubl., *Chaetolepis* (DC.) Miq.); intrageneric polyploidy and dysploidy (*Aciotis*, *Pilocosta* Almeda & Whiffin, and *Pterolepis* (DC.) Miq.); intrageneric constancy of high diploid numbers (*Heterocentron* Hook. & Arn. and *Monochaetum* (DC.) Naudin); and auto- or allotetraploidy followed by chromosome loss (*Tristemma* Juss. and *Heterotis* Benth.).

In addition to intrageneric polyploidy and dysploidy, *Aciotis* also showed intraspecific polyploidy (pers. obs.); for example in *A. polystachya* from Ecuador, a western collection (*Freire-Fierro 2835* from Puerto Quito) had n = 10, while an eastern collection (*Freire-Fierro 2798* from Puyo) had n = 20, a number reported for the first time for the species.

AGAMOSPERMY

Three Ecuadorian species of *Aciotis* were tested for agamospermy, which has been previously recorded in Melastomataceae (Renner, 1989, and references therein; Ramírez & Brito 1990; Goldenberg & Shepherd 1998). Flower development of pollinated and unpollinated flowers was observed in one or two populations of *A. polystachya*, *A. rubricaulis*, and *A. purpurascens*. None of the Ecuadorian populations studied showed evidence of agamospermy, but Renner (1989) reported agamospermy in a Brazilian population of *A. acuminifolia*.

The first population studied belongs to *A. polystachya* from the western side of the Andes (Pichincha: Puerto Quito, 700 m, *Freire-Fierro 2835*). In this population, the style emerged from the bud at ca. 4:30 am. Shortly thereafter in order to prevent pollination, styles were cut from flowers. At ca. 6:00 am the anthers were also evident and the petals were totally open. Pollinators began to visit the plants at approximately 6:00 am and finished their visits by 10:00 am. It was possible to recognize the flowers with uncut styles that had been visited by of the yellowish color of their pollinated stigmas. The flowers, with and without cut styles, started closing their petals around 10:00 am, and were totally closed by noon. When the flowers were pollinated, the corolla persisted and the ovary developed into a fruit after four to five days, but when the flowers were not pollinated, the ovary and corolla fell off after one to four days (Fig. 7). This scenario suggests that this population of *A. polystachya* is not agamospermous. Mature fruits remained on the plants for more than 12 days, after which they opened longitudinally, releasing their seeds. Of approximately 338 flowers studied (on average, 2.2 open flowers per day × 11 individuals × 14 days), none produced fruits after the styles had been cut off. Similar results were obtained in a second population from the eastern side of the Andes (Napo: Jatun Sacha, *Freire-Fierro 2844*). Of approximately 56 flowers observed (on average, 2 open flowers per day × 7 individuals × 4 days), none produced fruits.

Similar observations were performed on a population of *A. rubricaulis* (Pichincha: Puerto Quito, *Freire-Fierro 2834*). Flowers again were open between 6:00 and 10:00 a.m, after which time the petals started closing, and were totally closed around noon. If the flowers were not pollinated, the ovary fell off on the fourth day (Fig. 8), or when strong rain fell, the old unpollinated flowers fell by the morning of the second day. Of approximately 184 flowers studied (on average 1.2 open flower per day × 11 individuals × 14 days), none produced fruits after their styles had been cut. When the flowers were pollinated, they developed into fruits that remained on the plant for more than 10 days. Therefore, the studied population of *A. rubricaulis* was also not agamospermous.

The fourth population studied belonged to *A. purpurascens* and was located on the eastern side of the Andes (Napo: Jatun Sacha, *Freire-Fierro 2840*). The pattern of flowering was similar to that of *A. rubricaulis*. Of approximately 585 flowers (on average, 3 open flowers per day × 13 individuals × 15 days) whose styles were cut, none produced fruits (Fig. 9), which leads to the conclusion that *A. purpurascens* is also not agamospermous.

Although *A. rubricaulis* and *A. purpurascens* are not agamospermous, they can reproduce vegetatively by stolons.

HABITATS AND DISTRIBUTION

Aciotis is a Neotropical genus distributed from Mexico to Brazil and in the Lesser Antilles (Fig. 10). The species are present in dry or wet habitats from sea level up to 1700 m elevation.

Only *A. indecora* and *A. purpurascens* occur in Mexico, and only *A. rubricaulis* and *A. indecora* occur in Belize, Guatemala, Honduras, and Nicaragua. All three species are found in Costa Rica and Panama, as well as in South America.

Two South American species also occur in the Lesser Antilles, *Aciotis acuminifolia* (Martinique, Grenada) and *A. purpurascens* (Martinique). David Don based *Aciotis* on an Anderson collection of *A. purpurascens* (his "*Aciotis discolor*") from St. Vincent, the only record from that island. This collection was prepared from plants cultivated at the botanical garden on St. Vincent of which Anderson was superintendent from 1785 to 1811. Two specimens at BM, also attributed to Alexander Anderson from "West Antilles," may have been cultivated at that garden as well. Of the BM specimens *Anderson 287* is *A. annua*

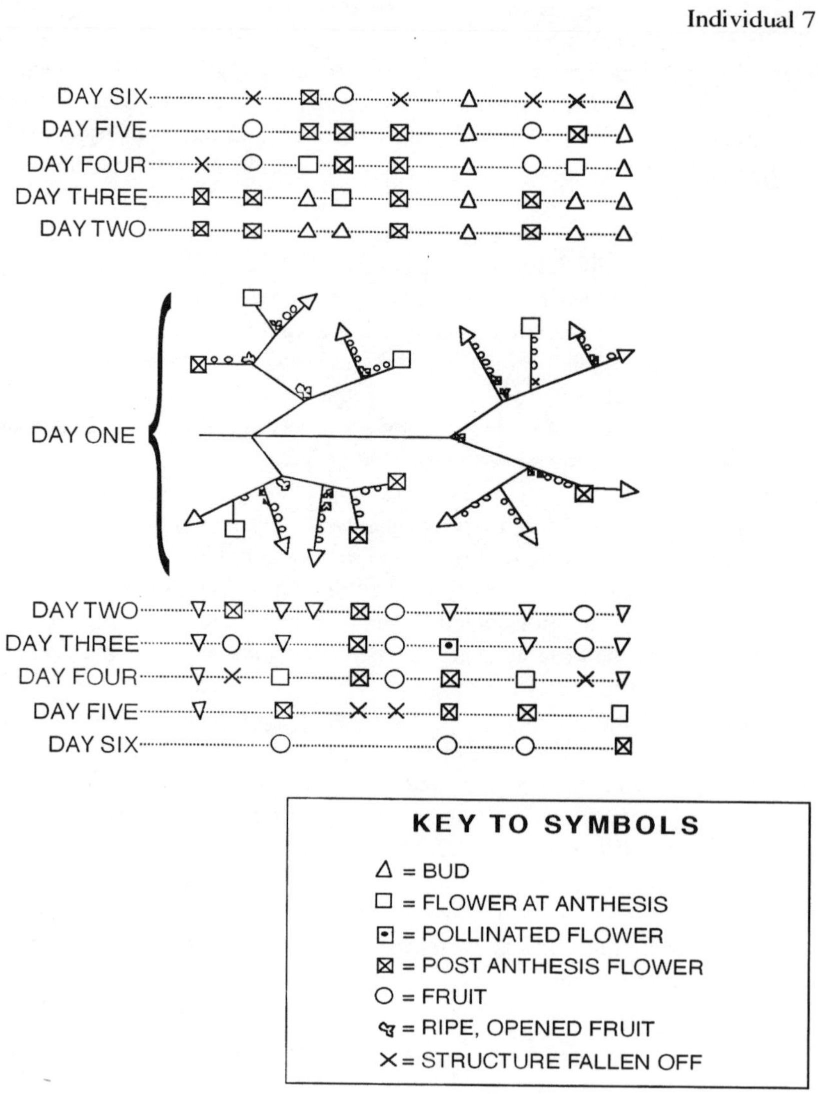

FIG. 7. Pattern of anthesis and post-anthesis in an individual of *Aciotis polystachya* from a population in western Ecuador.

and *Anderson s.n.* is *A. indecora*; neither species has otherwise been recorded from the West Indies. Alternatively, Anderson may have collected them during his visits to Guyana and Suriname (Lanjouw & Stafleu 1954).

In South America, *Aciotis* is most diverse in Colombia (10 species), Brazil (all 13 species), and Venezuela (9 species). Suriname has the fewest of species of *Aciotis* (5). *Aciotis acuminifolia*, *A. indecora,* and *A. purpurascens* are common to all South American countries, but *A. ferreirana*, *A. paludosa,* and *A. wurdackiana* are endemic to Brazil, where they have a restricted distribution. *Aciotis circaeifolia*, although occurring in Colombia, Guyana, and Brazil, has also a narrow range in these countries.

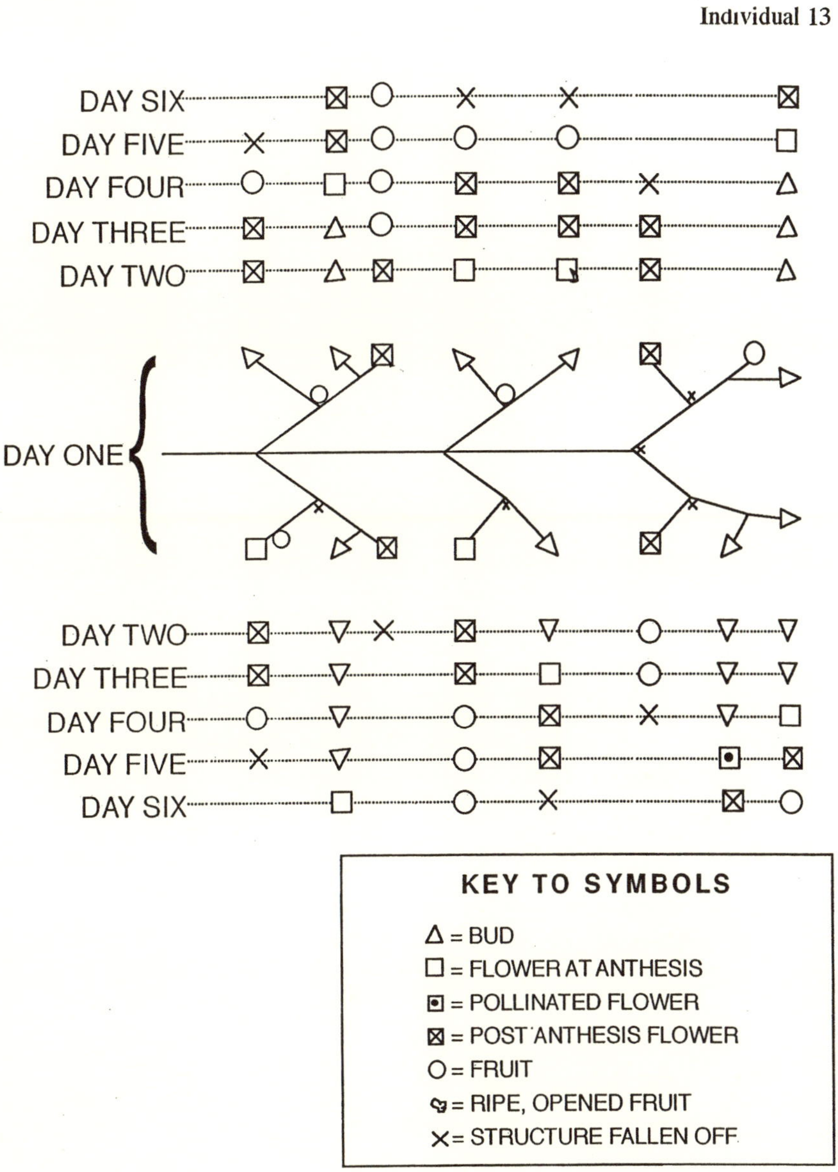

FIG. 8. Pattern of anthesis and post-anthesis in an individual *Aciotis rubricaulis* from a population in western Ecuador.

Glaziou collections present a special problem. As described by Wurdack (1970), some collections attributed to Glaziou are actually those of others and the locality information given on such labels is often erroneous. Brazilian material of *Aciotis* includes such spurious Glaziou collections. For example, according to the labels *Glaziou 9822*, *9817*, and *9877* were collected in São Fidelis (Rio de Janeiro) but in reality are from Amazonian Brazil. The first two were already cited by Cogniaux (1885) in *Flora*

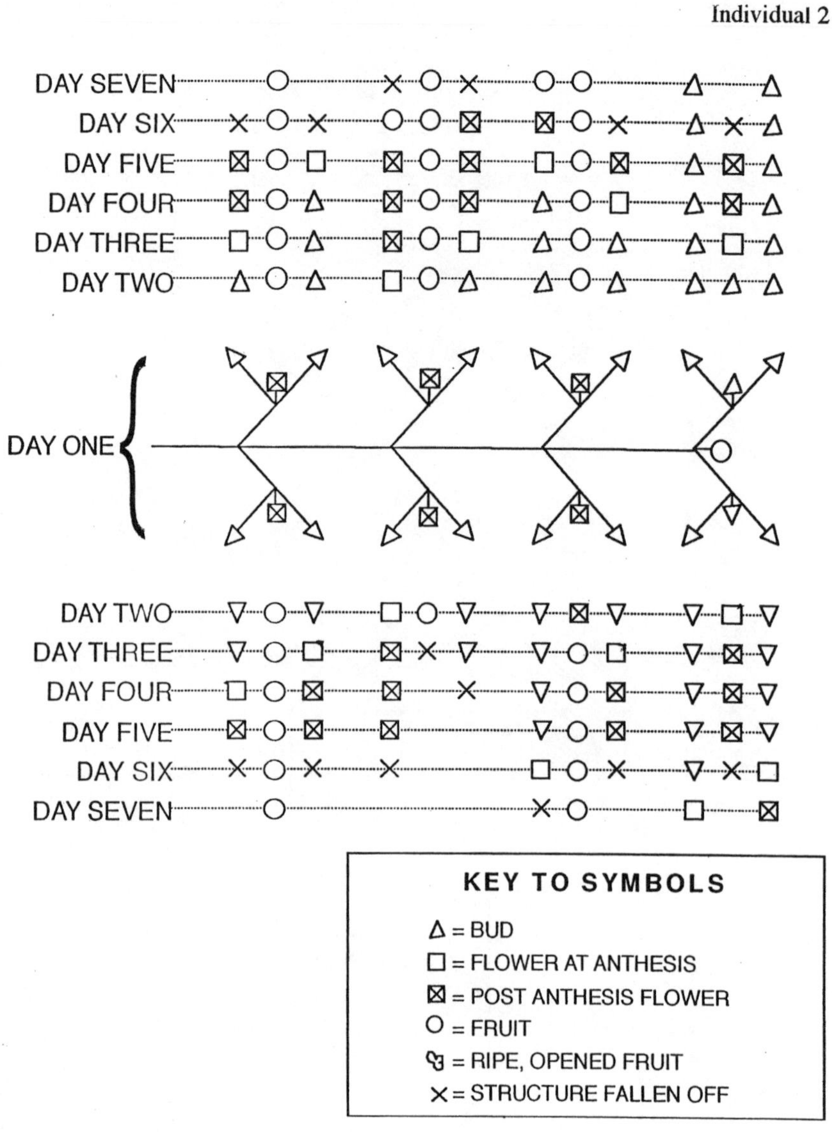

FIG. 9. Pattern of anthesis and post-anthesis in an individual of *Aciotis purpurascens* from a population in eastern Ecuador.

brasiliensis as *Schwacke 9817* (*A. annua*, cited under *A. dysophylla)* and *Schwacke 9822* (*A. ferreirana*, cited under *A. circaeifolia*). "*Glaziou 9877*" is *A. polystachya.* In order to avoid confusion I prefer not to cite any Glaziou collections in the specimens citations associated with the species treatments; however, for the sake of completeness, they are included in the "List of Numbered Collections Examined" that is presented at the end of this study.

FIG. 10. Distribution of *Aciotis*.

GENERIC AND INFRAGENERIC AFFINITIES

Students of Melastomataceae have placed *Aciotis* in Tibouchineae (Triana 1873; Cogniaux 1891), or in Melastomeae s.l. (Renner 1993; Clausing & Renner 2001) and traditionally have included species with capsular fruits and cochleate seeds with a tuberculate testa. The cochleate seed, also observed in *Aciotis*, was one of three synapomorphies recognized for this tribe in an early cladistic analysis of the family (Renner 1993). Whiffin and Tomb (1972) accept the placement of *Aciotis* in Tibouchineae [Melastomeae], although they mention that *Aciotis*, together with other genera of Melastomeae (*Acisanthera*, *Appendicularia* DC., *Comolia*, *Ernestia*, and *Nepsera* Naudin), has a microlicioid seed type, that is, an elongate to cochleate seed with a foveolate testa. The melastomoid genus *Siphanthera* Pohl has been suggested as close relative of *Aciotis* (F. Almeda, pers. comm.), and in a recent molecular analysis of Melastomataceae (Clausing & Renner 2001), *Nepsera* and *Aciotis* are placed as sister taxa (with a 100% bootstrap value).

Some species of *Aciotis* could be confused with *Leandra* and *Ossaea* (both in Miconieae) because of the cymose thyrsoid inflorescences and small fruits. This resemblance has even tricked experts of Melastomataceae. For example, Gleason (1935) described a specimen of *Aciotis indecora* as a new species, *Leandra montana*. Similarly, a specimen of *Aciotis ornata* [*Croat & Cogollo 52299* (MO)] was determined by Wurdack as *Ossaea robusta* and *O.* cf. *bracteata* in 1982, and as *Aciotis* cf. *aequatorialis* in 1984.

Species of *Aciotis* seem to be separable into two groups; however, several species bridge them. The first group, which includes *A. rubricaulis*, *A. circaeifolia*, *A. indecora*, and *A. purpurascens*, has thyrsoid to paniculate inflorescences and baccate fruits; the second group includes *Aciotis annua*, *A. ornata*, *A. polystachya,* and *A. wurdackiana*, and has filiform sclereids, biparous to double biparous cymose inflorescences, and dry capsular fruits. Yet, *A. acuminifolia*, *A. ferreirana*, *A. olivieriana*, *A. paludosa*, and *A. viscida* have characters of both the first and second groups (e.g., biparous to double biparous cymose inflorescences and fleshy capsules in *A. acuminifolia*; double biparous cymose inflorescences and baccate fruits in *A. ferreirana* and *A. paludosa*; double biparous cymose to thyrsoid inflorescences and capsular fruits in *A. olivieriana* and *A. viscida*).

Naudin (1850) and Cogniaux (1885, 1891) divided *Aciotis* into two groups based on anther morphology. The first group included species with elongate, linear anthers with pedoconnectives; and the second group included species with short, ovoid anthers lacking pedoconnectives. Krasser (1893) followed these authors; he recognized the first group as sect. *Euaciotis* and listed *A. rubricaulis* [as *A. alata*], *A. indecora,* and *A. purpurascens* as representatives. The second group he named sect. *Spennera*, and included *A. acuminifolia* [as *A. dichotoma*], *A. annua*, and *A. paludosa*. Yet, in this scheme as well almost half of the species cannot be placed in either section, and acceptance of a formal infrageneric classification would be artificial and impractical.

Within *Aciotis*, it may be suggested that the South American species with capsular fruits and filiform sclereids (*A. annua*, *A. wurdackiana*, *A. ornata,* and *A. polystachya*) are the more basal in the genus, and that species with baccate fruits but lacking filiform sclereids, and reaching Central America (*A. purpurascens*, *A. indecora,* and *A. rubricaulis*) are the more recently derived. Phylogenetic analyses using morphologic and molecular data are needed to resolve the relationships within *Aciotis*.

TAXONOMY

Aciotis D. Don, Mem. Wern. Nat. Hist. Soc. 4: 283, 300. 1823. *Aciotis* sect. *Euaciotis* Krasser in Engler & Prantl, Nat. Pflanzenfam. III. 7: 150. 1893.—Type: *Aciotis discolor* D. Don. [=*Aciotis purpurascens* (Aublet) Triana].

Spennera Martius ex DC., Prodr. 3: 115. 1828. *Aciotis* sect. *Spennera* (Martius ex DC.) Krasser, in Engler & Prantl, Nat. Pflanzenfam. III. 7: 150. 1893.—LECTOTYPE, here designated: *Spennera annua* Martius ex DC. [=*Aciotis annua* (Martius ex DC.) Triana].

Annual or perennial herbs with or without stolons. Stems with conspicuous or inconspicuous wings, glabrous or with glandular or eglandular trichomes. Leaves opposite, petiolate; leaf blade ovate to elliptic or oblong, apex acuminate to obtuse, base attenuate to cordate, membranous or papyraceous when dried. Inflorescence a double biparous or uniparous cyme, a panicle, or a thyrse, with one to seven pairs of paracladia of the first order, the hypopodia short or elongated (Fig. 2). Flower (Fig. 3) with a hypanthium; sepals 4, apically free and basally connate; petals 4, free, white to pinkish;

stamens 8, the anthers white or purple, pedoconnective present or absent (Fig. 5), dorsal or ventral projections absent; ovary half superior, bi- or rarely trilocular, apex glabrous or with glandular trichomes. Fruits (Fig. 6) capsular, the pericarp with or without filiform sclereids and caducous or persistent, the placenta caducous or persistent; or fruits baccate, without filiform sclereids, and always caducous when ripe. Base chromosome number: x = 10 (Almeda 1997a).

Key to the Species of Aciotis

Note: Sterile material cannot be identified with certainty.

1. Leaves abaxially with web-like striations formed by filiform sclereids (Fig. 1); ripe fruits greenish to whitish, not translucent.
    2. Inflorescences with the terminal paraclades bearing 5–8 flowers per cm; bracts 2–5 mm long.
        3. Young stems pubescent; leaves (0.6–) 1.4–2.6 (–4.5) cm wide, ovate to elliptic, the base truncate to subcordate. 1. *A. annua.*
        3. Young stems glabrescent; leaves 0.6–1.4 cm wide, elliptic, the base cuneate to truncate, never subcordate. 2. *A. wurdackiana.*
    2. Inflorescences with the terminal paraclades bearing 2–3 flowers per cm; bracts 0.8–1.8 mm long.
        4. Plants prostrate; young stems woolly, with reddish eglandular trichomes; leaf apex typically obtuse; bracts < 1.2 mm long. 3. *A. ornata.*
        4. Plants erect; young stems densely pubescent, with pinkish to whitish glandular trichomes; leaf apex acute to acuminate; bracts > 1.5 mm long. 4. *A. polystachya.*
1. Leaves without web-like striations underneath; ripe fruits externally greenish, in some species translucent.
    5. Inflorescences cymes or double biparous cymes, with 1–3 pairs of first-order paraclades.
        6. Leaf margin crenate-serrate, the base truncate. 5. *A. ferreirana.*
        6. Leaf margin serrulate, the base attenuate, cuneate, obtuse (sometimes truncate in *A. acuminifolia*), decurrent, or cordate.
            7. Young stems glabrous or pilose to ferruginous-woolly, usually with eglandular trichomes; inflorescences wider than long (3–6 cm long × 2–6.5 cm wide); ovary glabrous at apex.
                8. Young leaves woolly, with reddish trichomes, the apex acute to obtuse; pericarp and placenta caducous. 6. *A. paludosa.*
                8. Young leaves glabrous or pilose, the apex acute to acuminate; pericarp caducous, placenta persistent. 7. *A. acuminifolia.*
            7. Young stems glandular-pubescent; inflorescences longer than wide (6–14 cm long × 4–5.5 cm wide); ovary with glandular trichomes at apex. 8. *A. olivieriana.*
    5. Inflorescences thyrses or a panicles, with (2–) 4–8 pairs of first-order paraclades.
        9. Young stems glabrous (or strigose), or with trichomes > 1.5 mm long; petals abaxially glabrous; ovary glabrous at apex.
            10. Stems slightly quadrangular, wings inconspicuous; leaf base cordate; stems and leaves beset with scattered dark, ferruginous, glandular trichomes. 9. *A. circaeifolia.*
            10. Stems quadrangular, wings conspicuous; leaf base attenuate to cuneate (rarely cordate); stems and leaves glabrous (or strigose) or with dense pubescence of white to pinkish, usually eglandular trichomes.
                11. Stems and leaves glabrous or sparsely strigose; rachides and hypanthia glabrous. 10. *A. purpurascens.*
                11. Stems and leaves pilose; rachides and hypanthia with glandular or eglandular trichomes. 11. *A. rubricaulis.*
        9. Young stems densely pubescent, with glandular trichomes < 1 mm long; petals abaxially glandular-pubescent; ovary glandular-pubescent at apex.
            12. Inflorescences > 6 cm wide, first-order hypopodia > 1.3 cm long; ripe fruits baccate, the pericarp membranous, translucent, caducous. 12. *A. indecora.*
            12. Inflorescences < 5 cm wide, first-order hypopodia < 1.2 cm long; ripe fruits capsular, the pericarp coriaceous, not translucent, persistent. 13. *A. viscida.*

1. Aciotis annua (Martius ex DC.) Triana, Trans. Linn. Soc. London 28: 52. 1871. *Spennera annua* Martius ex DC., Prodr. 3: 115. 1828.—TYPE: BRAZIL. Piauí, *Martius s.n.* (holotype: G-DC, microfiche: MO!; isotypes: E! K! M!, photo of M isotype: F! G! GH! NY!).

Spennera dysophylla Bentham, J. Bot. (Hooker) 2: 296. 1840. *Aciotis dysophylla* (Bentham) Triana, Trans. Linn. Soc. London 28: 52. 1871.—TYPE: BRAZIL. Amazônas: "Barcelos on the Rio Negro," [00°58′S, 62°57′W], *R. H. Schomburgk I 932* (holotype: K!; isotypes: BM! E! G! GH! K! NY! W!).

Spennera latifolia Bentham, J. Bot. (Hooker) 2: 296. 1840, non *Spennera latifolia* DC., 1828.—TYPE: GUYANA. 1836, *R. H. Schomburgk I 16* (holotype: K!; isotype: BM!).

Spennera tetraptera Miquel, Comm. phytogr. 2: 76, t. 10, fig. E. 1840.—TYPE: SURINAME. "regione savane dicta," April, *Focke 162* (holotype: U). [Placement of name according to Wurdack et al. (1993)]

Herbs (2–) 15–20 cm tall, erect (rarely prostrate); stolons absent (rarely present); young stems quadrangular with conspicuous wings, glandular- or eglandular-pubescent with trichomes <1 mm long. Leaf blades (1–) 2.5–4.7 (–7.5) cm long, (0.6–) 1.4–2.6 (–4.5) cm wide, ovate to elliptic, apex acute, base truncate to subcordate (cuneate), margin serrulate, papyraceous, 5–7-veined; adaxially dark green, strigose, the trichomes eglandular, abaxially whitish green to light green, pubescent, strigose, the trichomes eglandular, with web-like striations; petiole 0.3–1.2 (–2.5) cm long, with narrow wings and pinkish trichomes <1 mm long. Inflorescence a double biparous or uniparous cyme, 5–7.5 (–13) cm long, 2–4.5 (–10.5) cm wide, with 2–3 (–5) pairs of first order paraclades; rachis with eglandular trichomes; internodes 1.2–1.7 cm long; first order hypopodia (0.3–) 1.2–2.5 cm long, biparous or uniparous; second order hypopodia (0.2–) 0.5–1.8 (–2) cm long, uniparous; third order hypopodia ca. 0.2 cm long, uniparous; fourth order hypopodia absent; terminal paraclade (0.5–) 1.5–3 (–4) cm long with 5–6 flowers per cm; bracts 2–2.5 mm long, 1–1.2 mm wide, ovate to deltate, apex acute, abaxially sericeous. Hypanthium 1.5–2.5 mm long, 0.8–1.5 mm in diameter, terete, with filiform sclereids in parenchyma, abaxially sericeous and glandular-pubescent, adaxially glabrous. Petals ca. 4 mm long, 1.5 mm wide, obovate, apex obtuse, base cuneate, with a terminal glandular trichome, abaxially glabrous. Stamens with filaments 2.5–3 mm long, anthers 1–1.5 mm long, purple, pedoconnective ca. 0.1 mm long. Ovary with filiform sclereids in parenchyma, apex glandular-pubescent. Fruit a capsule, pericarp and placenta persistent after dehiscence, 2.5–2.8 mm long, ca. 1.9 mm in diameter, bilocular, pericarp chartaceous, whitish, not translucent; seeds 0.25–0.35 mm long, 0.20–0.30 mm wide. Fig. 11.

Phenology. Flowering and fruiting throughout the year.

Distribution (Fig. 12). Southeastern Colombia, southern and eastern Venezuela, the Guianas, northern Brazil, and northern Bolivia; flooded areas and also in secondary and dryer habitats or sandy soils; 33–875 m.

Vernacular names. *Uzillabingoi* (Colombia: Amazonas; Witoto dialect); *boyuyo* and *garapata* (Venezuela: Amazonas); *nombró-poh* (Venezuela: Bolívar).

ADDITIONAL SPECIMENS EXAMINED. **Colombia.** AMAZONAS: Río Igara-Paraná (affl. Río Putumayo), corr. La Chorrera, territorio de indios Witoto Jitomagaro, familia Kuiru (17 km en aval de la Chorrera), *Gasche & Desplats 100* (G, US); La Chorrera, *Sastre 3467* (G, US).—ANTIOQUIA: Caracoly et Cisneros, *Mayor 255* (W).—VAUPÉS: basin de l'Amazone, Río Negro, a San Felipe, Comisaría del Vaupés, *Humbert 27398* (NY, US);

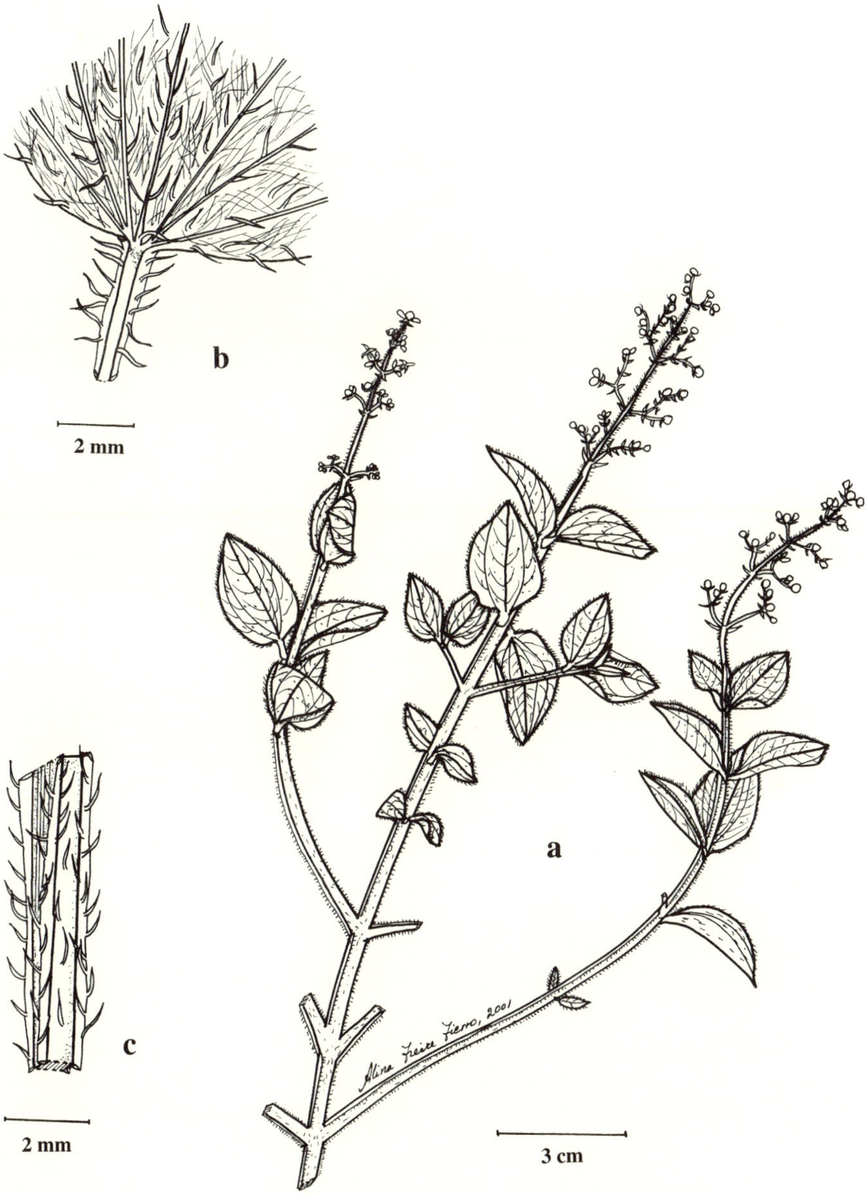

FIG. 11. *Aciotis annua*. a. Habit. b. Abaxial surface of leaves (the very fine lines represent the web-like striations). c. Portion of stem, showing the eglandular hairs borne marginally on the wings. (Based on: a, *Prance 30454*; b, c, *Bautista 136*.)

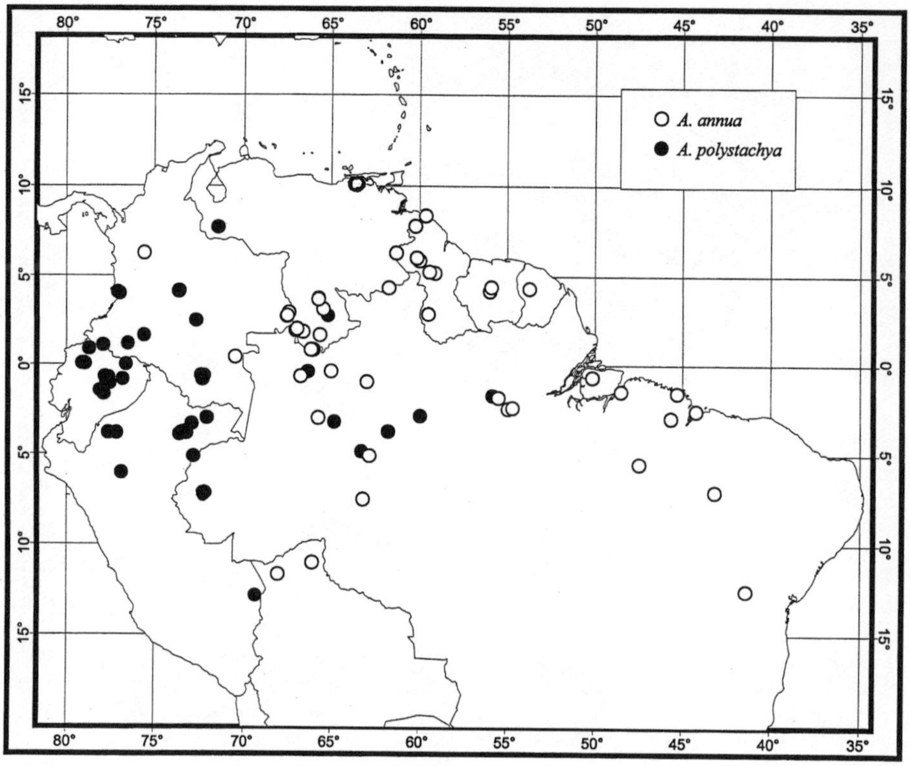

FIG. 12. Distribution of *Aciotis annua* and *A. polystachya*.

Río Pirparaná (tributary of Río Apaporis), Caño Teemeña, 00°25′N, 70°30′W, *Schultes & Cabrera 17204* (NY). **Venezuela.** AMAZONAS: Depto. Atabapo, Caño Iguapo (Aguas Negras), Alto Orinoco, 03°07′N, 75°28′W, *Aymard & Delgado 8211* (US); Cerro Neblina, *Croat 59584* (TEX); Río Negro, Mamurividi, laja along the Río Pasimoni, eastern bank, *Davidse & Miller 26707* (MO); Cerro Neblina, *Davidse & Miller 27283* (MO, NY, US); Cerro Neblina, Río Mawarinuma, vicinity of Puerto Chimo, 00°50′N, 66°05′W, *Gentry & Stein 46928* (MO, US); Dept. Casiquiare, Maroa, *Huber 2582* (US, VEN); Rio Negro, 4–7 km NE, *Liesner 7545* (MO); Cerro Neblina, *Liesner 17267* (US); Río Siapa, near base of Cerro Aracamuni, *Liesner & Carnevali 22762* (CAS, MO); Río Cunucunuma, Río Orinoco, *Maguire et al. 30400* (K, NY, US); Río Guainia, between Moroa and sabanita, 1 km E of Maroa, *Maguire et al. 36425* (F, K, NY, US, VEN); entre Caño Chola y Solano, 4 km S de Solano, *Morillo & Hasegawa 5009* (VEN); Carretera a San Carlos–Solano, 6–7 km S de Solano, 02°00′N, 66°58′W, *Morillo & Villa 5452* (VEN); selvas pluviales ribereñas del Río Casiquiare, entre la boca y la piedra Guachapita, *Stergios et al. 13112* (NY, US); Cerro Duida, base of SE slopes along Caño Negro, *Steyermark 58097* (F); Depto. Atabapo, norte del Cerro Duida, *Steyermark et al. 125799* (CAS); Río Pacimoni-Yatua, *Wurdack & Adderley 43416* (F, NY, RB, S, US).—BOLÍVAR: (Guayana Venezolana), región de los Ríos Icabarú, Hacha y cord. sin nombre, *Bernardi 2630* (G, NY); vicinity of Icarabú, 04°19′N, 61°44′W, *Croat 54107A* (MO, US); Mpio. Gran Sabana, sector El Pauji, *Fernández 1707* (VEN); Icarabú-Quebrada Negra, *Ruiz-Terán & López-Palacios 10962* (US); Depto. Piar, Zapata, a 10 km al SW de Icabarú, *Susach 1032* (AAU).—SUCRE: Campanito, Sabana de Piedra–Santa María, Dtto. Rivero, 10°25′N, 63°30′W, *Ruiz-Terán & López-Palacios 10078* (US). **Guyana.** Potaro-Siparuni, Mazaruni, *Bartlett s.n.* (K); between the Demerara and Berbice rivers, *de la Cruz 1573* (F, GH, MO, NY, UC, US); Upper Rupununi, near Dadanawa, 02°54′N, 59°00′W, *de la Cruz 1788* (F, GH, MO, NY, US, WIS); Upper Demerara River, *de la Cruz 2083* (F, GH, MO, NY, US); Waini River, Northwest district, *de la Cruz 3848* (F, GH, MO, NY, UC, US); Wanama River, NW district, *de la Cruz 3992* (NY, US); Rockstone, *Gleason 509* (GH, NY, US); Tukeit, *Graham 170* (K); East Coast Water Conservancy, SE of Georgetown, canal SE of Lamaha stop-off, *Hitchcock 16977* (GH, NY, S, US); E bank, Demerara River,

Atkinson Field, *Irwin 165* (TEX); Potaro River, below the Kaieteur, *Jenman 915* (K); Upper Demerara River, *Jenman 4013* (K); Tukeit, abandoned farm along Potaro river, 4 km below Kaieteur Falls, *Kvist et al. 209* (NY, US); Bartica, *Lang s.n.* (NY); Tukeit, 05°13′N, 59°27′W, *Maguire & Fanshawe 23046* (NY); Porknocker camp, end of Madhia Road, confluence of N Fork River and Minihaha, *McDowell et al. 4941* (NY); Bartica, *Potter 5395* (GH); Mazaruni, *Tutin 5* (BM); Amatuk, *Tutin 477* (BM, US); Potaro River, Kangaruna, *Whitton 56* (K). **Suriname.** Palisadeweg, *B. B. S. 272* (K); without locality, *Splitgerber s.n.* (BM, BR, K, W). **French Guiana.** St. Elie, 04°40′N, 54°35′W, *Lescure 264* (CAY). **Brazil.** AMAZONAS: Ramific. da Transamazônica, *Estrella & Bellusci 31-15376* (INPA); Camanáos, *Holt & Blake 585* (RB, NY, US); Mpio. Humaitá, 500 m ao Norte da BR-230 Km 2, 07°31′S, 63°10′W, *Janssen 242* (INPA, M); Marituba, Mpio. Belém, *Oliveira 6678* (MG); along Rio Marie at Manauna, *Poole 1879* (MG, MO, US, NY); Rio Xie, afluente do Rio Negro, *Silva et al. 38437* (INPA); Rio Negro, Santa Isabel, *Tate 77* (BM). —BAHIA: Andaraí, *de Carvalho et al. 1073* (K, US); Lençois, 12°34′S, 41°23′W, *Noblick 1784* (MBM, MO, US), antiga estrada de Barcellos, on the Rio Negro, *Trail 253* (K).—CEARÁ: without locality, *Gardner 932* (BM, G).—GOIÁS: Rod. Transamazônica, Mpio. Araguatins, *Hatschbach & Kummrow 38432* (MBM, US); Veredão do Relampago, duas leguas de Carolina, *Pires & Silva 2426* (NY).—MARANHÃO: Km 155 of BR-316, 03°00′S, 45°40′W, *Daly et al. D199* (NY); without locality, *Don 110* (BR); Mpio. Imperatriz, behind Rodobras station on S side of city of Imperatriz, *Eiten & Eiten 10219-B* (US); Island de São Luiz, estrada de Sacavem, *Froés 11518* (F, G, GH, MICH, MO, NY, S); without locality, *Gardner 6017* (BM, K, G, W); Rio Alto Turiaçu, Nova Esperança, *Jangoux & Bahia 225* (MG, NY).—PARÁ: Rivers Mojú and Acará, S of Belém, *Austin 4069* (MO); Monte Alegre, *Baker 133* (BM, BR, G, M, W); Benavides, estrada do Mosqueiro, PA-17, *Bautista 136* (MG); Belém, *Dahlgren & Sella 440* (F); Marajó, *G. Don, Jr. s.n.* (BM); 2 km S of Estrella do Norte, Mpio. Belém, *Drouet 2071* (F, GH, MICH, MO, NY, R, S, US); Monte Alegre, Serra de Parytuna, *Ducke MG-17154* (HB, US); Trombetas, Rio Jaramacarú, *Egler 256* (HB); Belém, 01°30′S, 48°30′W, *Guedes 2004* (R); Santa Isabel, *Huber 114* (BR, G, HB, R); Rio Capim, Cachoeira, *Huber 866* (G, HB, R, RB); Tapaná, *Killip & Smith 30367* (NY, US); near Antonio Lemos, *Krukoff et al. 5889a* (BM); campos de Ariramba, próximo do Rio Jaramacaru, *Martinelli et al. 12313* (INPA, MO, NY, US); Belém, *Pires & Black 322* (GH); Rio Arapiuns, Campo da Raposa, *Pires & Silva 338* (NY); bank of Rio Anajás just below mouth of Rio Guarajá, *Prance et al. 30454* (F, INPA, MO, NY, US); Península de Alter do Chão, *Sanaiothi s.n.* (INPA); São Joaquim, prope Pará, *Schwacke 170* (BR); Monte Alegre, *Schwacke 670* (RB); Belém, *Silva 92* (F, K, US); Acará, Jacarequara, Tapera, *Silva 548* (MG); Santarém, *Spruce 338* (US, W), *710* (K), *816* (W); Obidos, 01°52'S, 55°30'W, *Spruce s.n.* (K, W).—PIAUÍ: without locality, *Martius s.n.* (M). **Bolivia.** EL BENI: Vaca Diez, 3 km E of Riberalta, on road to Guayaramirím, then 2 km SE on side road, 11°00′S, 66°05′W, *Solomon 7687* (MO, US).—PANDO: Manuripi, 35 km al norte de Puerto America, 11°39′S, 68°11′W, *Jardim 2449* (USZ).

Aciotis annua can be recognized primarily by the dense inflorescences with conspicuous bracts and by the whitish color on the lower surface of the leaves, bracts, and fruits. The conspicuous wings on the stems and the 5–7-veined leaves with an obtuse to subcordate base are also distinctive. Most of the examined specimens have serrulate leaves, but the collection *Trail 253* (K) has doubly serrulate leaves. Several collections of *A. annua* had previously been identified as *A. polystachya*, perhaps because the leaf bases and vein numbers are similar; however, *A. polystachya* has wider and more lax inflorescences and much smaller bracts, and vegetative parts are covered with glandular trichomes.

Aciotis wurdackiana is similar to *A. annua* in the dense inflorescences, conspicuous bracts, and whitish coloring, but it differs in its elliptic leaves with cuneate to truncate base and glabrescent stems and leaves. *Aciotis annua* could be mistaken for the cordate-leaved representatives of *A. acuminifolia*, but that species is covered with glandular hairs and has lax inflorescences, fleshy capsules with caducous pericarp, and lacks filiform sclereids.

The Brazilian collection *Zerny s.n.* (W) has the winged stems and whitish fruits typical of *A. annua* but a lax inflorescence like that of *A. acuminifolia.* It was found in the area of Santarém, Pará, where both *A. annua* and *A. acuminifolia* occur and may represent a hybrid.

Renner (1984) reports that Brazilian populations of *A. annua* are pollinated by halictids and *Paratetrapedia* sp.

Whiffin and Tomb (1972) examined seeds of the Venezuelan collection *Wurdack & Adderley 43416*, deposited at US, for a study of seeds of Neotropical capsular-fruited species of Melastomataceae.

2. Aciotis wurdackiana Freire-Fierro, Novon 11: 166. 2001.—TYPE: BRAZIL. Pará: Rio Trombetas, vicinity of Cachoeira Porteira [01°55′S, 55°00′W], 21 May 1974, *Prance et al. 22220* (holotype: INPA!; isotypes: AAU! F! GH! K! MO! NY! S!).

Herbs ca. 40 cm tall, erect; stolons absent; young stems quadrangular with inconspicuous wings, initially with eglandular trichomes >1 mm long, becoming glabrescent. Leaf blades 1.2–3.1 cm long, 0.6–1.4 cm wide, elliptic, the base cuneate to truncate, the apex acute, the margin serrulate, papyraceous, 3–5-veined; adaxially dark green, pilosulous, the trichomes eglandular, abaxially whitish green, pubescent, the trichomes eglandular, with web-like striations; petiole 0.2–0.9 cm long, without wings, eglandular-pubescent, the trichomes >1 mm long. Inflorescence a cyme or a double biparous or uniparous cyme, 7.2–8 cm long, 2.2–2.5 (–4.5) cm wide, with 2–4 pairs of first order paraclades; rachis glandular-pubescent; internodes 0.8–1 cm long; first order hypopodia 0.2–0.3 cm long, biparous or uniparous; second order hypopodia 1.3–2.2 cm long, uniparous; third order hypopodia ca. 0.2 cm long, uniparous; fourth order hypopodia absent; terminal paraclade 0.6–1.2 cm long with 5–8 flowers per cm; bracts 2.5–5 mm long, 0.8–0.9 mm wide, narrowly elliptic, apex acute to acuminate, abaxially glandular-pubescent. Hypanthium 1.5–2.2 mm long, 1–1.2 mm in diameter, terete, with filiform sclereids in parenchyma, abaxially densely glandular-pubescent, adaxially glabrous or with few glandular trichomes. Petals ca. 2.5 mm long, 0.8–0.9 mm wide, narrowly elliptic, apex acute, base cuneate, without a terminal glandular trichome, abaxially glabrous. Stamens with filaments 2.2–2.5 mm long, anthers 1.2–1.8 mm long, purple, pedoconnective ca. 0.1 mm long. Ovary with filiform sclereids in parenchyma, the apex glandular-pubescent. Fruit a capsule, pericarp and placenta persistent after dehiscence, 1.5–2.2 mm long, 1–1.5 mm in diameter, bilocular, pericarp chartaceous, whitish, not translucent; seeds ca. 0.40 mm long, 0.20–0.30 mm wide. Fig. 13.

Phenology. Flowering and fruiting from May to July.

Distribution (Fig. 14). Known only from one area in Amazonian Brazil (Pará: Rio Trombetas, Cachoeira Porteira); disturbed areas along rivers, secondary forest, and wet areas in white sands; 10–760 m.

ADDITIONAL SPECIMENS EXAMINED. **Brazil.** PARÁ: Rio Trombetas, Cachoeira Porteira, prox. a foz do Rio Mapuera, 01°45′S, 55°50′W, *Campbell et al. P22526* (GH, INPA, K, NY, S, US), *Cid & Ramos 1038* (F, GH, INPA, MICH, NY, US), *Davidson & Martinelli 10364* (INPA, NY, US), *Martinelli 7039* (INPA, NY, US), *Silva & Santos 4708* (NY, US).

Aciotis wurdackiana is well marked by its very small leaves, dense inflorescences, and whitish capsular fruits. It is similar to *A. annua*, but differs in the narrower leaves and glabrescent stems. Collections of this species have sometimes been identified as *A. polystachya*. Although both species have filiform sclereids and capsular fruits, *A. polystachya* has obtuse to subcordate leaf bases, glandular-pubescent stems, and laxer inflorescences with terminal paraclades with only one to three flowers per cm. *Aciotis ornata* also has capsular fruits and filiform sclereids, but it differs in its obtuse leaf apices, laxer inflorescences, and shorter bracts (less than 1.5 mm long).

Aciotis wurdackiana may hybridize with *A. polystachya*; see Appendix 2.

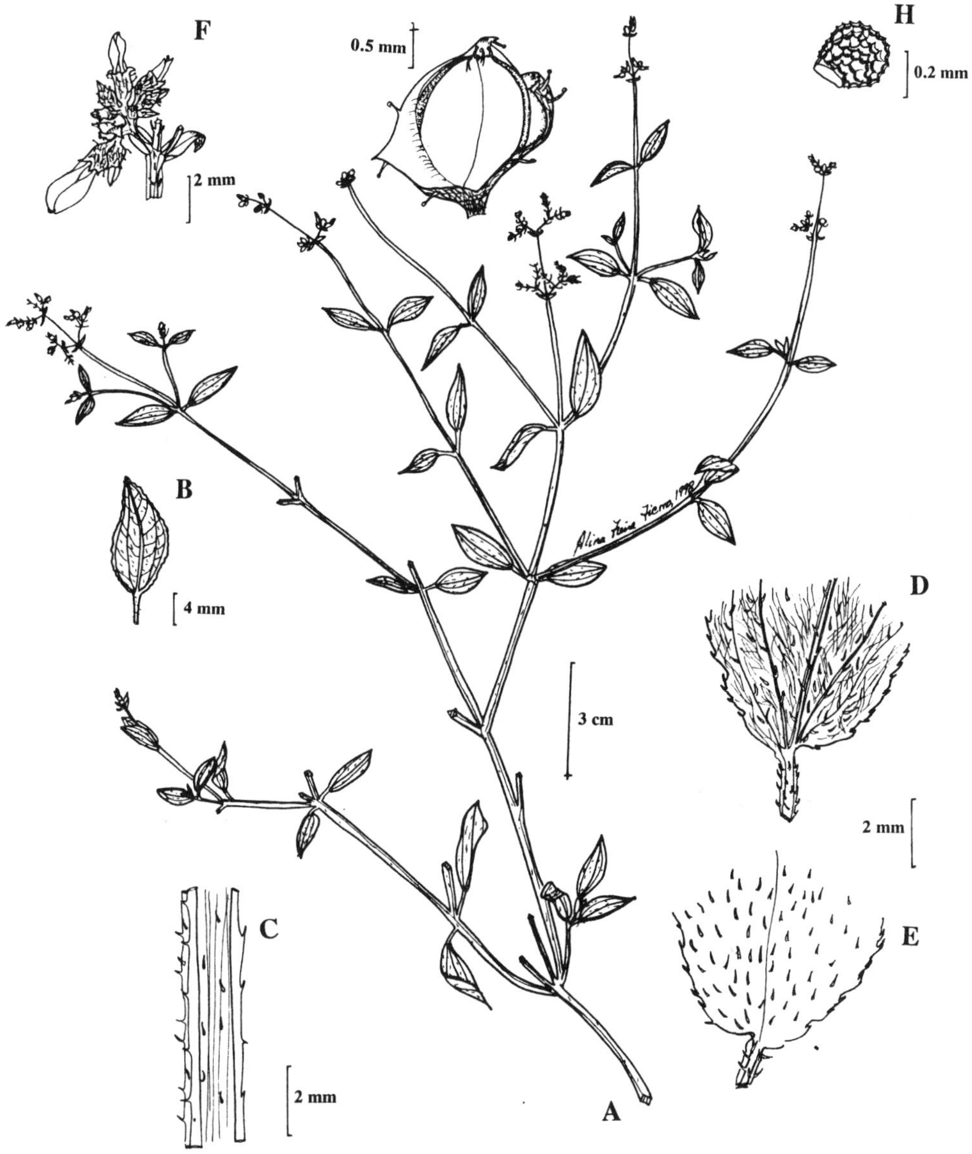

FIG. 13. *Aciotis wurdackiana*. A. Habit. B. Leaf. C. Stem. D. Abaxial surface of leaf (the very fine lines represent the web-like striations). E. Adaxial surface of leaf. F. Portion of inflorescence. G. Capsular fruit with glandular ovary apex. H. Seed with foveolate testa. (Based on: A–F, *Prance 22220*; G, H, *Campbell P22526*; reprinted from *Novon* with the permission of the Missouri Botanical Garden.)

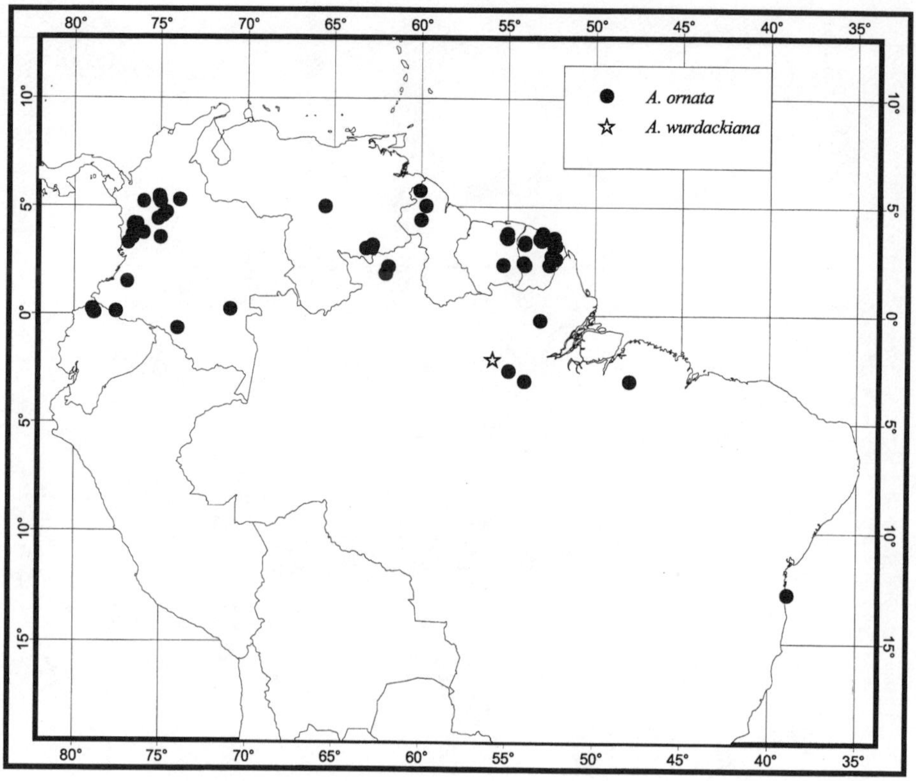

FIG. 14. Distribution of *Aciotis ornata* and *A. wurdackiana*.

3. Aciotis ornata (Miquel) Gleason, Brittonia 1: 149. 1932. *Spennera ornata* Miquel, Stirp. surinam. select. 56, t. 13. 1851.—TYPE: SURINAME. "prope Bergendaal" [Berg en dal, left Suriname River, ca. 05°15′N, 55°00′W], May, *Focke 1277* (holotype: U, digital image: MO!; isotypes: K! NY!).

Aciotis herbacea Cogniaux in Eichler, Fl. bras. 14(3): 468. 1885.—TYPE: SURINAME. "in sylvis humidis regionis interioris ad flum. Suriname," Oct, *Hostmann & Kappler 1832* (holotype: P, photos: F! NY!; isotype: S!)

Herbs 5–15 cm long, prostrate; stolons present; young stems quadrangular with inconspicuous wings, pubescent with woolly reddish eglandular trichomes <1 mm long. Leaf blades 2–7.5 (–9.2) cm long, 2.2–4 (–5.2) cm wide, ovate (rarely cordiform), apex obtuse or sometimes acute, base obtuse to subcordate, margin serrulate, papyraceous, 5–7-veined; adaxially green, pubescent with ferruginous or white eglandular trichomes, abaxially whitish green or purplish, pubescent with eglandular trichomes, with web-like striations; petiole 0.3–2 (–4) cm long, with narrow wings and woolly pubescence, the trichomes reddish, eglandular, <1 mm long. Inflorescence a cyme or a double biparous cyme, 5–8 cm long, 3–5.5 (–11) cm wide, with 1–3 pairs of first order paraclades; rachis with glandular trichomes; internodes (1–) 1.5–2.8 cm long; first order hypopodia 0.5–1.8 cm long, biparous or uniparous; second order hypopodia 0.2–0.4 cm long, biparous or uniparous; third order hypopodia 0.2–0.5 cm long, uniparous; fourth order hypopodia absent; terminal paraclade 1.5–4.5 (–6) cm long with 2–3 flowers per cm; bracts 0.8–1.2 mm long, 0.3–0.8 (–1.5) mm wide, ovate, apex acute, abaxially glabrous. Hypanthium 1.5–1.8 mm long, ca. 1.5 mm in diameter, angled, with filiform sclereids in parenchyma, abaxially densely glandular-pubescent, adaxially glabrous. Petals 1.8–2 mm long, 1–1.2 mm wide, obovate, apex obtuse, base attenuate, with a terminal glandular trichome, abaxially glabrous. Stamens with filaments 1.5–1.7 mm long, anthers 0.7–1 mm long, white or purplish, pedoconnective absent. Ovary with filiform sclereids in parenchyma, apex glandular-pubescent. Fruit a capsule, pericarp and placenta persistent after dehiscence, ca. 2.5 mm long, 1.8–2.5 mm in diameter, bilocular, pericarp coriaceous, whitish, not translucent; seeds 0.20–0.40 mm long, 0.15–0.35 mm wide. Fig. 15.

Phenology. Flowering and fruiting throughout the year.

Distribution (Fig. 14). Northwestern Ecuador and western and southeastern Colombia to central and southeastern Venezuela, the Guianas, and the eastern Brazilian Amazon, one record from coastal Bahia, Brazil; secondary areas and shaded understory habitats; 10–760 m.

Vernacular names. *Chichu paispa cum* (Colombia: Nariño); *mortiño* (Colombia: Chocó); *mape-remedio-tape*, *atoma-bé-remedio-tapé* (Ecuador: Esmeraldas).

ADDITIONAL SPECIMENS EXAMINED. **Colombia.** AMAZONAS: Río Mirití-Paraná, afluente del Río Caquetá, 00°05′S, 74°00′W, *Palacios & Jaramillo 400* (COL).—ANTIOQUIA: Virginias, 40 km E of Pto. Berrio, 06°24′N, 74°39′W, *Archer 1432* (US); Cáceres, El Doce, Bajo Cauca, Quebrada Puri, 215 km de Medellín, *Callejas & Atehortua 318* (COL, HUA, NY); Tarazá, corregimiento El 12, 200 km NO de Medellín, Km 6 del camino El-12, Barroblanco, 07°N, 76°W, *Callejas 1080* (NY); Mpio. Anorí, Corregimiento Providencia, Valle del R. Anorí, entre Dos Bocas y Anorí, Carretera a Algibes, 07°N, 75°W, *Fonnegra et al. 653* (HUA); Mpio. Sonsón, Río Verde, Hac. La Soledad, parte alta de Quebrada Curubital, 05°00′N, 75°00′W, *Gutiérrez 35585* (F, G, UC); Mpio. San Carlos, correg. El Jordán, Embalse de Punchina, quebrada La Villa, 06°13′N, 74°52′W, *MacDougal et al. 4180* (CAS, MO).—CAUCA: El Tambo, La Romelia, Parque Nac. Natural Munchique, Km 85–91, 02°32′N, 76°59′W, *Lozano & Ruiz 7082* (COL).—CHOCÓ: along road between Quibdó and Medellín, Km 175 E of Tutunendo, 05°45′N, 76°23′W, *Croat 56305* (MO); along road between Quibdó and Bolívar, between Km 175 and 176, 117–118 km E of Quibdó, 05°44′N, 76°28′W, *Croat 57504* (CAS, MO); Carretera Quibdó–Guayabal, *Espina et al. 1500* (COL, MO); Hoya del Río San Juan, alrededores de Noanama, 04°42′N, 76°55′W, *Forero et al. 4555* (COL, MO, US); Hoya del R. San Juan, Itsmina, 05°10′N, 76°41′W, *Forero & R. Jaramillo 5376* (MO, US); Hoya del R. Atrato, Beté, Orillas de Río Atrato, *Forero et al. 8938* (COL, MO); R. San Juan, Estribaciones del Cerro de la Mojarra, *Forero et al. 9511* (COL, MO); new road S. of Yuto to Lloro, *Gentry & Fallen 17840* (COL, MO, NY, US); Pizarro, *von Sneidern 4966* (A, LL, S); confluence of R. Catrú and R. Dubaso, *Warner & White 75* (COL, GH, MO).—NARIÑO: Mpio. Tumaco, Km 80, ca. 00°52′N, 77°38′W, *Mora 4200* (COL).—SANTANDER: vicinity of Barranca Bermeja, Magdalena Valley, between Sogamoso and Colorado Rivers, 07°06′N, 73°54′W, *Haught 1314* (BM, NY, US).—VALLE DEL CAUCA: Río Raposo, colinas la NE del acampamento de la Rockefeller, *Fernández 6055* (COL).—VAUPÉS: R. Vaupés, Mitú, alrededores, ca. 01°N, 71°W, *García-Barriga 15174* (COL). **Ecuador.** ESMERALDAS: Río Zapallo Grande (tributario R. Cayapa), 00°48′N, 78°52′W, *Barfod 41016* (AAU, QCA, QCNE); Playa de Oro 01°N, 79°W, *Thomas 10* (K). **Venezuela.** AMAZONAS: Ríos Pacimoni-Yatua, Casiquiare, along Río Yatua above mouth of Río Yacibo, *Maguire et al. 36536* (NY, US).—BOLÍVAR: Mpio. Raúl Leoni, 85 km por aire al S de Urinam, 04°36′N, 62°44′W, *Fernández 3488* (VEN); Serranía Maigualid Frente 9, HP 3, E de Río Caur Bosques ribereños y de islas del Río Zariapo, Convenio UNELLEZ-CVG (TECMIN), 06°45′N, 65°31′W, *Stergios & Delgado 13763* (US); Río Diamante Negro, tributary of Río Guaña, *Steyermark et al. 107106* (VEN, US); Expedición Alto Paragua, Raudales de Maihia a lo largo del Río Paragua, 04°25′N, 63°07′W, *Steyermark 90533* (US); Distrito Heres, selva en cabeceras del Río Paramichi, sitio Paramichi, al E del Cerro Marutaní, cerca de frontera Venezolano-Brasileira, 03°00′N, 62°00′W, *Steyermark 124274* (MO, NY).—DELTA AMACURO: San Víctor, past Quebrada Piedrader, Sierra Imataca, *Steyermark 87182* (NY, US). **Guyana.** Imoti Creek, Waini River, *Becket 8535* (K); Yarikita portage, N.W.D., *Fanshawe 5179* (NY, K); Cuyuni-Mazaruni, Aurora, 180 km upstream of Cuyuni River, 06°47′N, 59°44′W, *Guillespie & Tiwari 2072* (NY, US); 200 yds S of Chodikar landing, Acarai foothill,

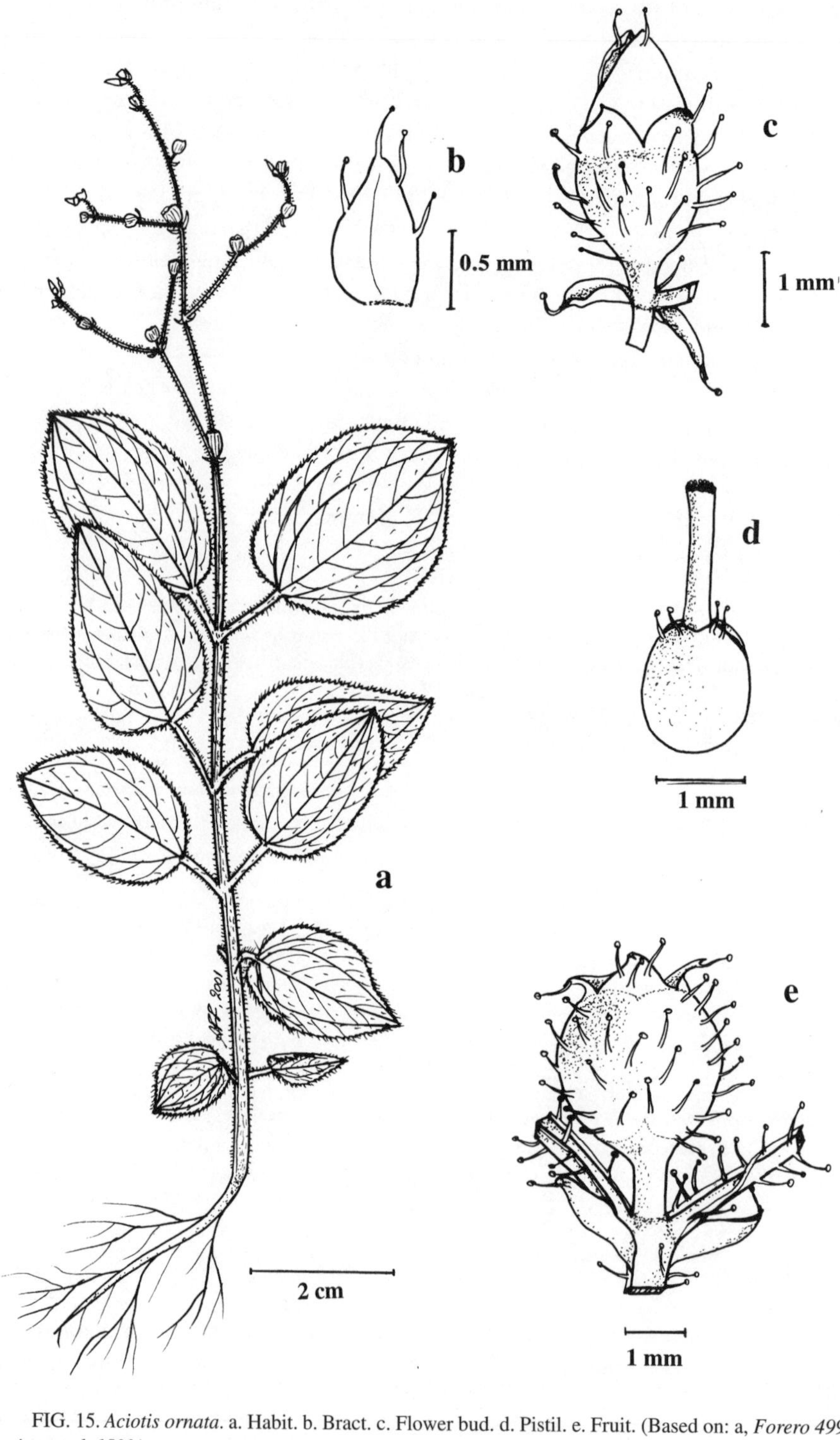

FIG. 15. *Aciotis ornata*. a. Habit. b. Bract. c. Flower bud. d. Pistil. e. Fruit. (Based on: a, *Forero 4993*; b–e, *Espina et al. 1500.*)

Guppy 7465 (K, NY, S, US); Barima-Waini Region, Barima River Head, Eclipse Falls, 5 minutes W of Arakaka, 07°38′N, 60°04′W, *Pipoly & Lall 8332* (NY, US). **Suriname.** Brokopondo, Baling, 05°N, 55°W, *Budelman 1041* (US); Jodensavanne-Mapane Kreek area (Suriname R.), 05°05′N, 55°04′W, *Lindeman 4220* (F). **French Guiana.** Montagne Caco, S de Cayenne, 04°30′N, 52°20′W, *Cremers 6929* (CAY, US); region de Paul Isnard, 04°40′N, 54°W, *Cremers 7919* (BR, CAY, US); Bicade, rive gauche du Grand Inini, 03°30′N, 55°15′W, *de Granville 612* (CAY, COL, US); Crique Gregoire, echantillon a la Grande Chute, 05°05′N, 53°W, *de Granville B.4364* (CAY, US); Crique Baton Pilon, rive gauche de l'Oyapock, 03°50′N, 52°15′W, *de Granville T.1056* (CAY, US); route Nationale RN-2, PK-49, 04°35′N, 52°20′W, *Feuillet 3623* (CAY, US); Bourg de Maripasoula-Bassin du Maroni, 03°37′N, 54°05′W, *Fleury 590* (CAY); Ile Bagatelle-Basin du Sinnamary, 05°02′N, 53°02′W, *Hoff et al. 6375* (CAY); Approuague, sur la crique Ipoucin, au camp Salvat, 04°N, 52°30′W, *Oldeman B-2191* (CAY); au bord de la Sinnamary, 20 km du bourg, 05°15′N, 53°W, *Prévost I-543* (CAY, US); Bakaweko, 03°30′N, 54°W, *Sauvain 362* (CAY, US). **Brazil.** AMAPÁ: near cachoeiras das Guaribas, 00°24′N, 53°07′W, *Egler & Irwin 46439a* (MG).—AMAZONAS: Manaus–Caracarai Road, Km 148, *Bisby et al. P18110* (NY, US); margem direita do Río Tocantins, ca. 03°S, 48°W, *Silva da et al. 309* (INPA).—BAHIA: Ilhéus, Castelo Novo, 14°37′S, 39°12′W, *E. P. Velloso 1096* (RB).—PARÁ: Mpio. Oriminaná, Río Paru do Oeste, margem esquerda, ao lado da cachoeira Pancada, *Cid Ferreira et al. 2053* (NY, GH, INPA, US); Río Tapajós, Perto da Frunas, ca. 03°S, 54°W, *Duke s.n.* (RB); BR 163 Cuiabá–Santarém, Km 1129, vicinity of Igarapé Natal, 02°26′S, 54°55′W, *Prance et al. P25484* (F, MG, MO, NY, RB, S, US).—RORAIMA: SEMA Ecological Reserve, Ilha de Maracá, Roraima, 03°23′N, 61°51′W, *Milliken 533* (K); estrada Manaus–Caracarai, Km 329, Army post N of Wimari-Atoari Indian Reserve, Manaus–Venezuela Highway, *Steward et al. 33* (INPA, NY).

Aciotis ornata can be distinguished by its prostrate habit, the whitish color of its lower leaf surface (due to the web-like striations), the lax cymose to double cymose inflorescence, and the persistent placentas. Most of the Guyanan and Venezuelan representatives have leaves with an obtuse apex, but those from Colombia and Ecuador specimens have the leaf apex acute to obtuse.

Although the species is easily recognized, several collections have been determined as *A. aequatorialis* Cogn. [=*A. acuminifolia*] or as *A. polystachya*. Like most of specimens of *A. acuminifolia*, *A. ornata* has the stems densely eglandular-pubescent, cymes or double biparous cymes, and capsular fruits with evident placentas; however, *A. acuminifolia* differs in that the leaves have an acute apex and a truncate to attenuate-decurrent base, and lack filiform sclereids.

The inflorescence type and the leaf base are somewhat similar in *A. ornata* and *A. polystachya*; however, *A. polystachya* is an erect herb with densely glandular-pubescent stems, leaves with an acute to acuminate apex, and longer bracts.

Aciotis ornata is reportedly used to treat snakebites by the Cayapa Indians of Esmeraldas, Ecuador (*Barfod 41016*, *Kvist 48316*), who place crushed leaves directly on the bitten area. In Colombia, cooked leaves, called *Chichu paispa cum* (*González 37*), are one of the ingredients of a potion used against a "contra" [a curse].

4. Aciotis polystachya (Bonpland) Triana, Trans. Linn. Soc. London 28: 52. 1871. *Rhexia polystachya* Bonpland, Monogr. Melast. 2: 98, pl. 38. 1818. *Spennera polystachya* (Bonpland) DC., Prodr. 3: 116. 1828.—TYPE: BRAZIL. "in sylvis humidis Brasiliae, " *Bonpland s.n.* (holotype: P, photos: F! NY!).

Herbs 15–50 cm tall, erect; stolons absent; young stems quadrangular with inconspicuous wings, glandular-pubescent with dense pinkish to whitish trichomes <1 mm long. Leaf blades (1.5–) 2–6 (–7.8) cm long, 1–3 (–4) cm wide, cordiform to ovate, apex acute to acuminate, base obtuse to subcordate, margin serrulate, papyraceous, 5–7-veined; adaxially dark green, glandular-pubescent with dark reddish or whitish trichomes, abaxially whitish green, glandular-pubescent with dark reddish or whitish trichomes, with web-like striations; petiole (0.2–) 0.7–1.8 (–3) cm long, with narrow wings, pinkish glandular-pubescent with trichomes <1 mm long. Inflorescence a double biparous cyme, 4–8 cm long, 3–6 cm wide, with 2–3 (–4) pairs of first order paraclades; rachis glandular-pubescent; internodes (1–) 1.5–3.2 cm long; first order hypopodia 0.7–1.7 cm long, biparous; second order hypopodia (0.3–) 0.5–0.7 (–1) cm long, biparous or uniparous; third order hypopodia ca. 0.5 cm long, uniparous; fourth order hypopodia absent; terminal paraclade 1.5–3.5 (–4.5) cm long with 3 flowers per cm; bracts 1.5–1.8 mm long, 0.5–0.8 mm wide, ovate, apex acute, abaxially glandular-pubescent. Hypanthium 1.5–2 mm long, 1.5–2 mm in diameter, terete, with filiform sclereids in parenchyma, abaxially densely glandular-pubescent, adaxially glabrous or with few glandular trichomes. Petals 1–2.5 mm long, 0.6–1 mm wide, ovate, apex acute, base cuneate, with a terminal glandular trichome, abaxially with glandular trichomes. Stamens with filaments 1.8–3 mm long, anthers 1–1.8 mm long, purple, pedoconnective absent to ca. 0.1 mm long. Ovary with filiform sclereids in parenchyma, the apex glandular-pubescent. Fruit a capsule, pericarp and placenta persistent after dehiscence, 2.2–3 mm long, 1.5–2 mm in diameter, bilocular or sometimes trilocular, pericarp chartaceous, greenish, not translucent; seeds 0.30–0.35 mm long, ca. 0.25 mm wide. Figs. 3e, f, 4c, 5c, d, 6c, 16.

Phenology. Flowering and fruiting throughout the year.

Distribution (Fig. 12). Colombia, Ecuador, Peru, Venezuela, and the Brazilian Amazon Basin; primarily in secondary dry or swampy areas; 25–1200 m.

Vernacular names. *Charapa sacha*, *chirapa sacha*, *muchaquilla*, *mullaca colorada*, *mulaquillilla*, and *rifarillo* (Peru: Loreto); *sirichá* (Venezuela: Bolívar).

ADDITIONAL SPECIMENS EXAMINED. **Colombia.** AMAZONAS: Río Caquetá, Isla San Pablo, La Pedrera, *Jaramillo et al. 8090* (COL); Caño del Engaño, afluente del Río Caquetá, *Palacios & Jaramillo 278* (COL); Río Igara-Paraná (affl. Río Putumayo), corr. La Chorrera, *Sastre 3096* (G); Río Apaporis, entre el Río Pacoa y Río Kananarí, Soratama, *Schultes & Cabrera 12521* (GH, NY); Araracuara, Río Caquetá, ca. desembocadura del Yarí, 00°37′S, 72°23′W, *Torres et al. 3069* (COL).—CAQUETÁ: Solano, Araraquara, Pista de aterrizaje, 00°50′S, 72°19′W, *Arbeláez & Sueroke 761* (MO); Km 8 entre Belén de los Andaquies y San José del Fragua, *Bernal & Téllez 416* (COL); Cordillera Oriental, vertiente oriental, Sucre, 04°09′N, 73°37′W, *Cuatrecasas 9050* (COL, US); Florencia, Escuela Normal de Señoritas, *Echeverry 1659* (HUA); Morelia, *von Sneidern A.1021* (A, COL, LL, MICH, NY, S,).—CHOCÓ: Tadó, Corregimiento Playa de Oro, Rd to Santa Cecilia, *Espina & Arias 1171* (MO); Corregimiento de Marmolejo, Km 15 adelante de Guarató (Risaralda) en la carretera Santa Cecilia–Tadó, *Fernández et al. 8769* (COL, HUA, US); Tutunendo, 04°09′N, 73°37′W, *Uribe 3112* (COL).—META: Mpio. Villavicencio, Reserva Bavaria, ca. 04°09′N, 73°37′W, *Lozano et al. 7437* (COL).—NARIÑO: San Miguel, Tuquerres–Tumaco road, valley of Río Güisa, 01°04′N, 77°58′W, *Gentry et al. 20758* (MO).—PUTUMAYO: Mpio. Villa Garzón, carretera a Puerto Asís, 01°10′N, 76°34′W, *Fernández et al. 11387* (HUA, MO); Río Mocoa, en Puerto Limón, *Uribe 6579* (COL, NY).—RISARALDA: Mpio. Mistrató, corregimiento de Puerto de Oro, Vereda Chirrincha, orillas del Río Aguita, *Fernández et al. 8901* (COL, HB, HUA).—VALLE DEL CAUCA: Buenaventura, Bajo Calima, ca. 15 km N of Buenaventura, new Dindo area, 03°59′N, 77°02′W, *Gentry et al. 56712* (MO); vicinity of Bajo Calima, along road past Pulpapel Headquarters (at Km 17) to Bahía de Málaga, 04°03′N, 77°08′W, *Croat & Watt 70191* (MO, US).—VAUPÉS: Guaviare, Mpio. San José del Guaviare, Resguardo indígena de la comunidad Tucano de Panoré, a 5 min. del centro Sinchi, 02°28′N, 72°41′W, *López & Giraldo-C. 602* (MO). **Venezuela.** AMAZONAS: Alto Orinoco, Raudal de los Guaharibos, *Croizat 360* (F, NY); Raudal Montserrat, *Croizat 702* (NY); Atabapo, Cerro Duida, base on N side opposite Culebra, 03°44′N, 65°44′W, *Liesner 24537* (MO, US); San Carlos de Río Negro, *Renner 2001* (AAU, CAS, US); Depto. Atures, carretera Pto. Ayacucho–Samariago, *Stergios 3213* (AAU); Depto. Atabapo, Río Cunucunuma, entre las comunidades de Culebra y Huachamacari, entre el Cerro Duida y Huachamacari, 03°40′N, 65°45′W, *Steyermark et al. 125671* (NY); bosque a lo largo del Río Ocamo, vecindades de la Misión de Sta. María de los Guaicas, 02°45′N, 65°12′W, *Steyermark 106168* (MO, NY, US).—BOLÍVAR: selva de galería y sabana a lo largo del Río Kanarakuni, N and NW de Misión de Campamento Sanidad del Río Kanarakuni, *Steyermark 98226* (US).—MÉRIDA: Río Caparo, 31 km ESE of Sta. Bárbara, ca. 07°41′N, 71°28′W, *Liesner & González 9286* (MO, US, VEN). **Ecuador.** ESMERALDAS: Playa de Oro, estero y cascada de San Juan, 00°52′N, 78°45′W, *Alvarez et al. 784* (QCNE).—NAPO:

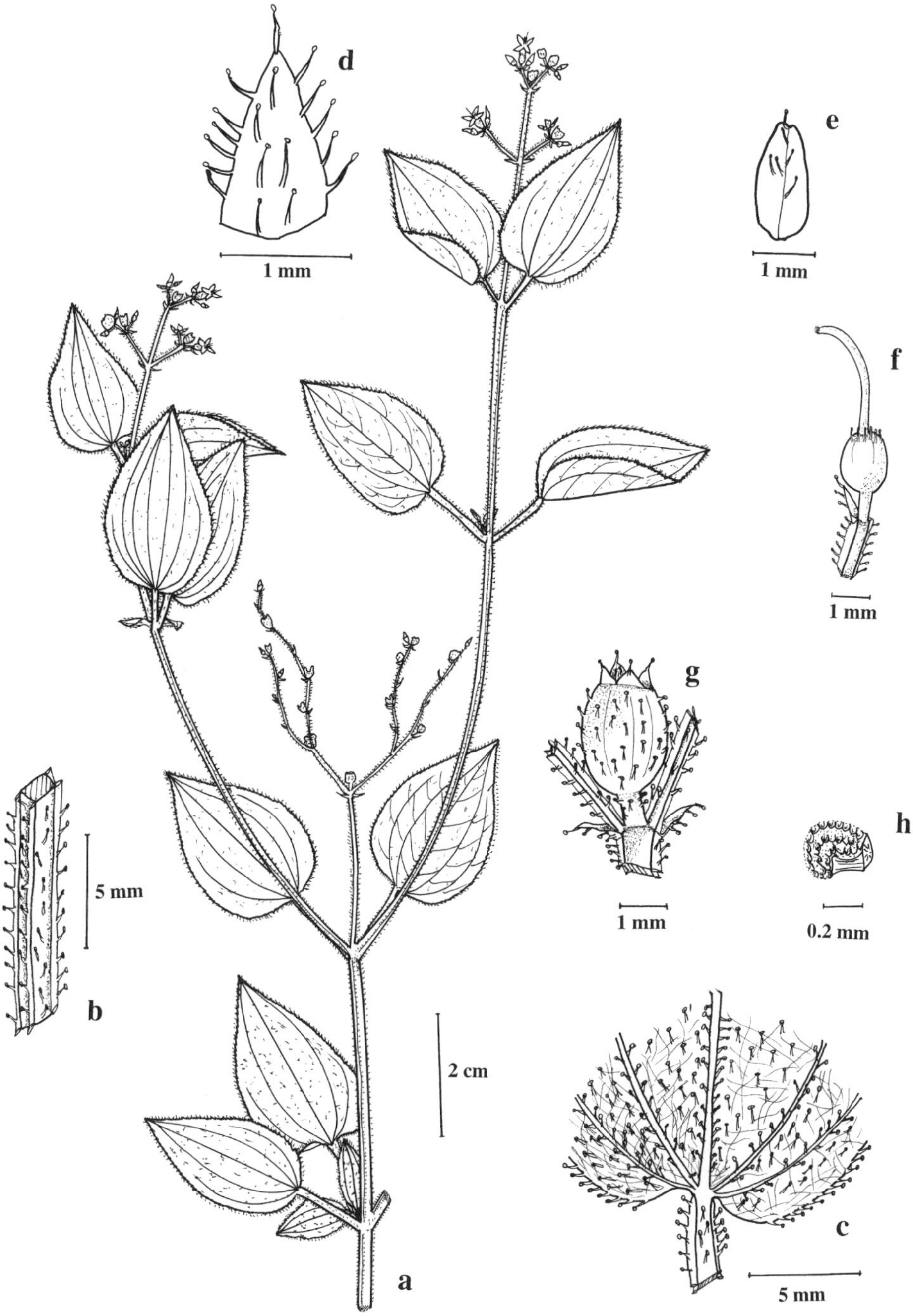

FIG. 16. *Aciotis polystachya*. a. Habit. b. Portion of stem. c. Abaxial leaf surface near leaf base (the very fine lines represent the web-like striations) d. Bract. e. Abaxial view of petal. f. Ovary and style; glandular trichomes present on the ovary apex. g. Fruit. h. Seed. (Based on: a, *McDaniel & Persinos 10694*; b–h, *Freire-Fierro 2769b*.)

Coca–Archidona, Km 30, vía Hollín–Loreto, ca. 00°45′S, 77°40′W, *Freire-Fierro & León 2750* (MO, QCA); Jatun Sacha 01°04′S, 77°36′W, *Freire-Fierro 2844* (QCA); Puerto Napo–Puyo rd, Km 3, 01°03′S, 77°47′W, *Freire-Fierro & León 2756* (MO, QCA).—PASTAZA: Mera, 01°28′S, 78°08′W, *Asplund 18327* (S); Puyo–El Porvenir–Puyopungo rd, ca. 01°40′S, 77°55′W, *Freire-Fierro 2811* (MO, QCA); Río Chullana, ca. 15 km N of Sarayacu, 01°44′S, 77°29′W, *Lugo 4146* (AAU, GB, MO, US).—PICHINCHA: Carretera Quito–Pto. Quito, Km 113, 10 km al N de la carretera principal, Reserva ENDESA, Corporación Forestal Juan Manuel Durini, 00°05′N, 79°02′W, *Betancourt 200* (GB, US); Los Bancos–Esmeraldas, Km 3, 00°02′N, 78°59′W, *Freire-Fierro 2769b* (MO, QCA); Pto. Quito, ENDESA 00°05′N, 79°02′W, *Freire-Fierro 2827a, 2835* (MO, QCA).—SUCUMBÍOS: old road Lago Agrio–San Miguel, Km 8, ca. 00°00′N, 76°40′W, *Freire-Fierro & León 2744* (MO, QCA). **Peru.** AMAZONAS: Condorcanqui, valle del Río Santiago, Quebrada Caterpiza, 2–3 km atrás de la comunidad de Caterpiza, 03°50′S, 77°40′W, *Tunqui 459* (MO).—LORETO: Iquitos, *Asplund 14067* (S); Maynas, Iquitos, Quistococha, 03°51′S, 73°13′W, *Ayala 640* (F, MO, NY); Bora Native community of Brillo Nuevo, Yaguasyacu River (affluent of Ampiyacu river), ca. 150 km ENE of Iquitos, 03°00′S, 72°05′W, *Denslow & Treacy 35* (F, MICH, US); Maynas, Mishana, trail from village to Campamento 1, 03°50′S, 73°30′W, *Gentry et al. 28922* (MO); Yurimaguas, *Ule 6267* (G); Maynas, Dtto. Amazonas, Quebrada Sucusari; 03°20′S, 72°55′W, *Vásquez & Jaramillo 15935* (MO); Pto. Almendras; Río Nanay, 03°45′S, 73°15′W, *Vásquez & Jaramillo 1182* (MO, NY); La Victoria on the Amazon River, Peru/Brazil, *Williams 2978* (F, NY).—MADRE DE DIOS: Tambopata, alrededores del Huerto Amazónico, 12°49′S, 69°18′W, *Reynel et al. 5241* (MO); along banks of Río La Torre at end of Swamp Trail, Explorer's Inn, near the confluence of Río Tambopata and Río La Torre, 39 km SW of Puerto Maldonado, 12°50′S, 69°20′W, *Smith et al. 740* (US).—SAN MARTÍN: Moyobamba, 06°04′S, 76°56′W, *Mathews 1280* (E, K); Lamas, near Tarapoto, *Williams 6362* (F, NY); Rioja, *Woytkowski 6077* (GH, MO). **Brazil.** ACRE: Río Branco, *Calderón & Soderstrom 2313* (INPA, K, US); Cruzeiro do Sul, Río Jurua and Río Moa, Río Moa at Cachoeira Grande, 07°39′S, 72°41′W, *Prance et al. 12519* (C, F, INPA, M, MICH, NY, RB, S, US).—AMAZONAS: Rio Javari, margem direita 6 horas depois de Atalaia do Norte, Palmarí, 05°08′S, 72°49′W, *Braga et al. 3147* (INPA); Cachoeira Caranguejo, Río Cauabury, *Holt & Blake 421* (US); Manaus, 02°53′S, 59°58′W, *Killip & Smith 30085* (US); Foz do Jutaly, Solimões, *Kuhlmann 1215* (BM); Manaus–Caracarai rd, Km 45, *Renner 214* (INPA, US).—PARÁ: Balterra, beira do R. Tapajós, Pindobal, *Black 471824* (RB); Rio Trombetas, entre Lago Jacaré e Cachoeira Porteira; 01°45′S, 55°50′W, *Martinelli 7001* (INPA, NY, US); Camanaus–Uaupés, ca. 00°23′S, 66°20′W, *Prance et al. 15686* (C, F, INPA, IJ, M, MG, MICH, MO, NY, R, S, US).—RORAIMA: base of Serra Tepequem, 03°45′S, 61°45′W, *Prance et al. 4268* (COL, F, G, INPA, MG, NY, RB, S, US).

Aciotis polystachya is easily recognized by the erect habit, the dense glandular pubescence covering the entire plant, and the cordiform to ovate leaves. It could be confused with cordate-leaved specimens of *A. acuminifolia*, but it differs by its filiform sclereids in the mesophyll and its glandular-pubescent petals and ovary apex. Another species with similar foliage and pubescence is *A. indecora*; however, that species has thyrsoid to paniculate inflorescences and baccate and caducous fruits.

Aciotis polystachya may hybridize with *A. acuminifolia*, *A. rubricaulis*, and *A. wurdackiana*; see Appendix 2.

Renner (1984, 1989) reports that Brazilian populations of *A. polystachya* are pollinated by species of *Paratetrapedia* Moure 1941, *Augochlora* Smith 1853, and *Euglossa* Latreille 1802 and by *Augochloris callichroa*. In Ecuador, I observed visits by Colletidae and Megachilidae bees.

In Peru, *A. polystachya* (as "*muchaquilla*") is used to treat tuberculosis and kidney disease. For its preparation (*Tina 2317*), one teaspoon of ground leaf material is put in a glass of hot water and covered for 15 minutes, then the filtered solution is drunk twice a day for 10 days. The species (as "*mullaquillilla*"; *Tello & Tina 2180*) is also used for kidney ailments.

5. Aciotis ferreirana Brade, Inst. Nac. Pesq. Amazônica Publ. Bot. 8: 10, fig. 2. 1958.—TYPE: BRAZIL. Mato Grosso: B. A. M, Igarapé de São Raimundo [12°49′S, 57°24′W], 21 Oct 1955, *Chagas s.n. =INPA 2198* (holotype: INPA!; isotype: INPA!).

Herbs 12–20 cm tall, erect; stolons absent; young stems quadrangular with inconspicuous wings, glabrous to glandular-pubescent with trichomes <0.5 mm long. Leaf blades 3.5–4.8 cm long, (1.5–) 2–3.3 cm wide, ovate, apex acuminate, base truncate, margin crenate-serrate, membranous, 3–5-veined; adaxially green, glabrous, abaxially light green, glabrous, without web-like striations; petiole 1.3–2 cm long, with narrow wings, glabrous to glandular-pubescent with trichomes <1 mm long. Inflorescence a double biparous cyme, (3.5–) 4–7 cm long, 2–2.5 cm wide, with 3 pairs of first order paraclades; rachis glabrous or with glandular trichomes; internodes 1–1.5 cm long; first order hypopodia 0.2–1 cm long, biparous; second order hypopodia 0.1–0.2 cm long, biparous or uniparous; third order hypopodia absent to ca. 0.1 cm long, uniparous; fourth order hypopodia absent; terminal paraclade ca. 0.2 cm long, with 1 flower per cm; bracts 1–1.5 mm long, ca. 0.5 mm wide, ovate, apex acute, abaxially glabrous. Hypanthium 2.5–2.6 mm long, 1.7–1.8 mm in diameter, terete, without filiform sclereids in parenchyma, glabrous or with a few glandular trichomes on the costae, adaxially glabrous. Petals ca. 3.1 mm long, ca. 1.5 mm wide, oblong, apex acute, base truncate-cuneate, without a terminal glandular trichome, abaxially glabrous. Stamens with filaments ca. 2.8 mm long, anthers ca. 1.7 mm long, pinkish, pedoconnective ca. 0.1 mm long. Ovary without filiform sclereids in parenchyma, apex glabrous. Fruit baccate, pericarp and placenta caducous when ripe, 2.8–3 mm long, 2.8–3.2 mm in diameter, bilocular, pericarp membranous, green when ripe, translucent; seeds ca. 0.50 mm long, ca. 0.30 mm wide. Fig. 17.

Phenology. Collected in flower in October and in fruit in November.

Distribution (Fig. 18). Endemic to western Brazil; usually in humid places.

ADDITIONAL SPECIMENS EXAMINED. **Brazil.** AMAZONAS: "Amazonia," 8 Aug 1876, *Schwacke 9822* [labeled *Glaziou 9822*] (BR, C, R); Mpio. Humaitá, Rio Madeira, banks of Rio Ipixuna, 07°33′S, 63°01′W, *Prance et al. 3335* (INPA, NY).—RONDÔNIA: Rio Ouro Preto, afluente do Pacca Nova, 10°45′S, 64°15′W, *Kuhlmann 444* (RB).

Aciotis ferreirana is easily identifiable by the crenate-serrulate, ovate leaves, narrow wings on the stems, dense inflorescence, and baccate fruits. *Aciotis annua* and *A. wurdackiana* also have dense inflorescences; however, *A. ferreirana* has glabrous leaves, the ovary apex glabrous, and lacks filiform sclereids.

Cogniaux misidentified *Schwacke 9822* as *A. circaeifolia*, which differs from *A. ferreirana* in the serrulate and cordate leaves, the thyrsoid inflorescence longer than 7 cm, and the fruit longer than 4 mm.

Although known from an area visited by several collectors, *A. ferreirana* has been collected only four times. Perhaps the species has specific ecological requirements that restrict it to narrowly defined habitats.

6. Aciotis paludosa (Martius ex DC.) Triana, Trans. Linn. Soc. London 28: 51. 1871.

Spennera paludosa Martius ex DC., Prodr. 3: 115. 1828.—TYPE: BRAZIL. Rio de Janeiro: Sebastianópolis [22°54′S, 43°10′W], 1817, *Martius s.n.* (holotype: G-DC, microfiche: MO!; isotypes: G! M!, photos of M isotype: F! G! GH! NY!).

Rhexia herbacea Raddi, Mem. Mat. Fis. Soc. Ital. Sci, Modena, Pt. Mem. Fis. 20: 127, pl. 1, fig. 5. 1828.—TYPE: BRAZIL. Rio de Janeiro: "cresce nei prati e nei boschi inondati di Mandiocca, specialmente attorno i piccoli ruscelli," 1817–1818, *Raddi s.n.* (holotype: FI, photos: GH! MO! MICH! NY!).

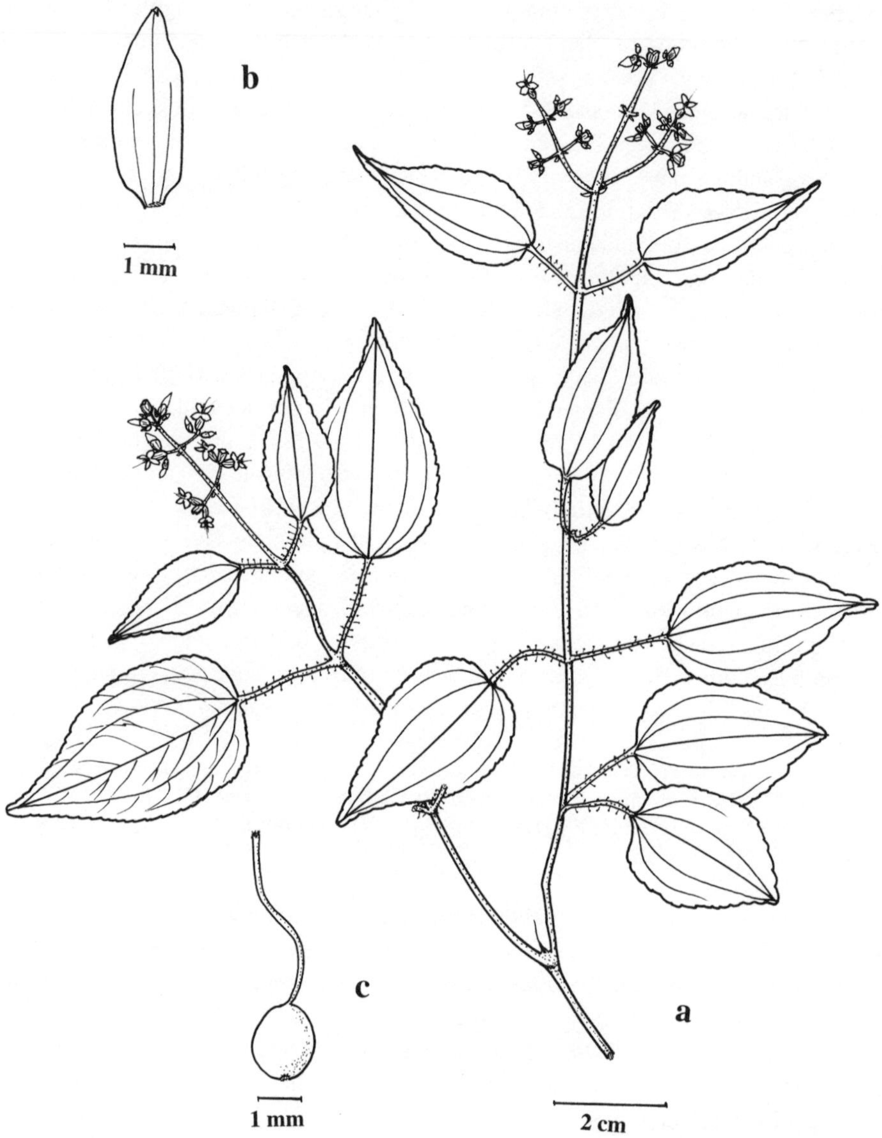

FIG. 17. *Aciotis ferreirana*. a. Habit. b. Abaxial view of petal. c. Ovary and style. (Based on *Chagas s.n.* = *INPA 2198*.)

Spennera brachybotrya DC., Prodr. 3: 115. 1828. *Aciotis brachybotrya* (DC.) Triana, Trans. Linn. Soc. London 28: 51. 1871.—TYPE: BRAZIL. Without locality, *Martius* (lectotype, here designated: G-DC, microfiche: MO!; isolectotype: BR!).

Spennera latifolia DC., Prodr. 3: 115. 1828, non *Spennera latifolia* Bentham, 1840. *Aciotis paludosa* var. *latifolia* (DC.) Triana, Trans. Linn. Soc. London 28: 51. 1871. *Aciotis latifolia* (DC.) Cogniaux in Eichler, Fl. bras. 14(3): 464. 1885.—TYPE: BRAZIL. Bahia: Ilhéus [13°01′S, 40°01′W], 1816, *Wied-Neuwied s.n.* (holotype: B, destroyed, photos: F! GH! MO! NY!; lectotype, here designated: BR!).

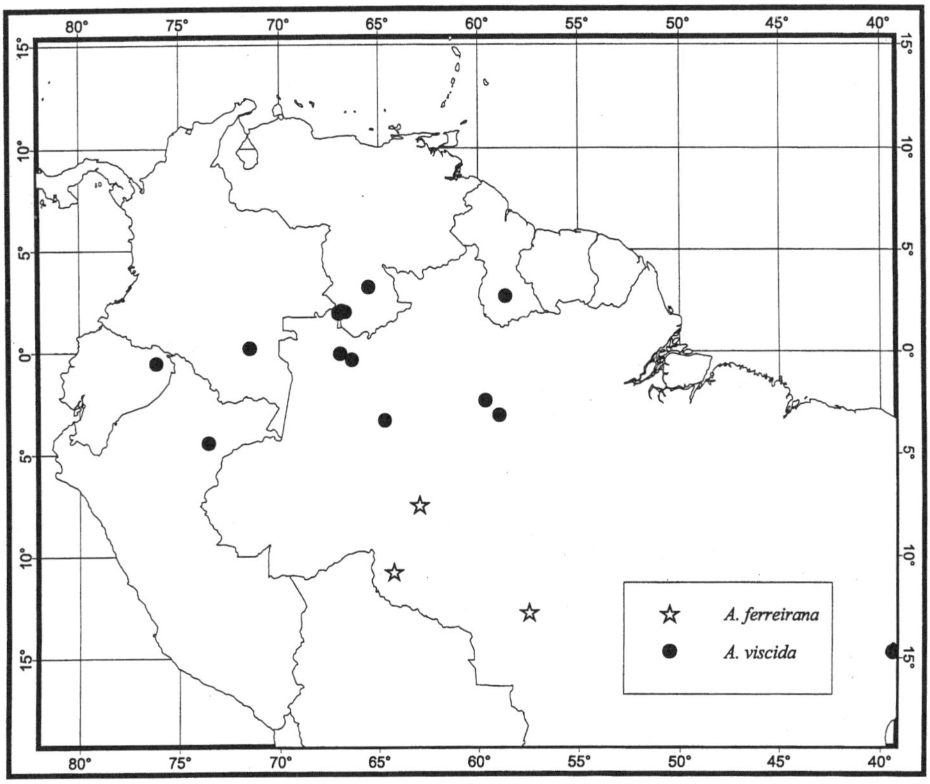

FIG. 18. Distribution of *Aciotis ferreirana* and *A. viscida*.

Spennera ferruginea Naudin, Ann. Sci. Nat. Bot., sér. 3, 14: 143. 1850. *Aciotis ferruginea* (Naudin) Triana, Trans. Linn. Soc. London 28: 51. 1871. *Aciotis brachybotrya* var. *ferruginea* (Naudin) Cogniaux in Eichler, Fl. bras. 14(3): 466. 1885.—TYPE: BRAZIL. "Amérique équatoriale," *Bonpland 49* (holotype: P, photos: F! MO! NY!, fragment: F!).

Aciotis martiana Cogniaux in Eichler, Fl. bras. 14(3): 463, pl. 104, fig. 2. 1885.—TYPE: BRAZIL. Bahia: Ilhéus [14°46′S, 39°04′W], *Martius 490* (holotype and fragment: BR!; isotypes: K! M! NY!).

Aciotis spiritusanctensis Brade, Arch. Jard. Bot. Rio de Janeiro 16: 11, pl. 6. 1959.—TYPE: BRAZIL. Espirito Santo: Norte do Rio Doce [21°06′S, 41°31′W], Sep 1949, *Jose Mendonça* & *Jair Vieira 28* (holotype: RB, photos: F! GH! NY!).

Herbs 0.12–0.20 m tall, usually erect, rarely prostrate; stolons absent (rarely present); young stems quadrangular with conspicuous wings, woolly, beset with reddish eglandular trichomes >1 mm long. Leaf blades (2–) 3.5–10 (13.5) cm long, (1–) 1.5–6 (–7) cm wide, elliptic to ovate, apex acute to obtuse, base attenuate to cuneate (rarely obtuse), margin serrulate, membranous, (3–) 5-veined; when young woolly, beset with reddish trichomes on both surfaces; when mature adaxially dark green, glabrescent, abaxially light green, pubescent to glabrescent, without web-like striations; petiole (0.5–) 0.9–3 (–4) cm long, with narrow wings, woolly, beset with reddish trichomes >1 mm long. Inflorescence a double biparous (rarely uniparous) cyme, 3–4.5 (–10) cm long, 2–6.5 (–10.5) cm wide, with (1–) 2 (–3) pairs of first order paraclades; rachis with few long glandular trichomes; internodes 0.6–1 (–2.5) cm long; first order hypopodia 0.5–1.2 cm long, biparous; second order hypopodia 0.5–1 cm long, biparous or uniparous; third order hypopodia 0.3 (–1) cm long, uniparous; fourth order hypopodia absent to 0.2 mm long, uniparous; terminal paraclade 1.5–3 cm long with 2–3 (–4) flowers per cm; bracts 1.2–1.5 (–3) mm long, 0.5 (–1) mm wide, ovate, apex aristate, abaxially glabrous. Hypanthium ca. 3 mm long, 2–3 mm in diameter, angled, without filiform sclereids in parenchyma, abaxially with few glandular trichomes, adaxially glabrous. Petals 2.5–5.5 mm long, 1.5–2 mm wide, oblong, apex acute, base cuneate, with or without a terminal glandular trichome, abaxially glabrous. Stamens with filaments ca. 1.8 mm long, anthers 0.8–1 mm long, white, pedoconnective absent to ca. 0.1 mm long. Ovary without filiform sclereids in parenchyma, apex glabrous. Fruit baccate, pericarp and placenta caducous when ripe, 2–3 (–4) mm long, 2–3 (–3.5) mm in diameter, bilocular, pericarp membranous, translucent; seeds 0.50–0.60 mm long, 0.30–0.40 mm wide. Fig. 19.

Phenology. Flowering and fruiting in all months, except June (i.e., mid-winter).

Distribution (Fig. 20). Southern Brazil; wet and swampy habitats, in "Mata Atlântica" [=Atlantic forests]; 50–700 m.

ADDITIONAL SPECIMENS EXAMINED. **Brazil.** BAHIA: margem da Rodovia Itabuna–Ilhéus, *Belém 1639* (US); Itabuna, Eunapolis, Itabela, Itamaraju, *dos Santos 918* (NY, RB, US); Camacã–Rio Branco Rd, *dos Santos 1434* (US); Ba. Uruçuca, Faz. São Jorge, *dos Santos 643* (US); Porto Seguro, Fazenda Pedro Deiró, *Duarte 6017* (HB, RB, US); entre Bom Gosto e Olivença, *Fróes de Lemos 20019* (NY); Rio Pau Velho, Mpio. Nova Viçoa), *Hatschbach & Silva 50740* (US, MBM); Mpio. Caravelas, Rodovia BR-418, a 51 km da BR-101, *Mori et al. 10568* (US); Mpio. Wenceslau Guimarães, ca. 3 km W of Nova Esperança, W edge or Reserva Estadual Wenceslau Guimarães, 13°36′S, 39°43′W, *Thomas et al. 9243* (NY, US); Ilhéus, Pirataquirrá, 14°46′S, 39°4′W, *Velloso 1035* (R); R. P. Seguro a Eunapolis, *Pinheiro 1751* (US).—DISTRITO FEDERAL: Brasilia, matas do Jardim Botânico, 22°43′S, 51°24′W, *Pereira s.n.* (HB).—ESPIRITO SANTO: Cachoeira do Itapemirim, Vargem Alta, *Brade 19749* (NY, RB); Norte, Serra de Cima, Mpio. Nova Venecia, *Duarte & Gomes 3736* (HB, RB); Lagoa Juparanã (Mpio. Linhares), ca. 12°S, 40°W, *Hatschbach & Cervi 51366* (MBM, US, Z); Lagoa Juparanã, Mpio. Linhares, *Hatschbach et al. 58032* (MBM, W); Mata Fria, Mpio. Venda Nova do Imigrante, *Hatschbach et al. 65213* (US); Reserva Forestal da CVRD, Linhares, entrada da Reserva, *Menandro 280* (US); Macuco, Reserva de Sooretama, *Sucre 5644* (RB).—MATO GROSSO: Barra do Garças, 15°52′S, 52°15′W, *Sidney & Onishi 1108* (US); Camararé, 12°15′S, 58°55′W, *Hoehne 4262* (BM, GH).—MINAS GERAIS: Parque Estadual do Rio Doce, Marlieria, trilha do Vinhático, *Costa s.n.* (BHCB); Rio S. Antonio, Mpio. Conceição do Mato Dentro, *Hatschbach s.n.* (MBM); Fazenda da Barra, Piáu, *Heringer 1726* (RB); Gortocazes, Rio Doce, *Kuhlmann 6543* (RB); Morro de Pilar, Km 2, *Shepherd 10206* (K).—PARANÁ: Rio Guaraqueçaba, Mpio. Guaraqueçaba, 24°S, 49°W, *Hatschbach 26249* (AAU, C, MBM, MO, S, UC, US); Socavão, perto do Bananal, *Manduça Penha s.n.* (R). —RIO DE JANEIRO: Mpio. Campos, Lagoa de Cima, *Araujo & Maciel 4045* (GUA); Tinguá, *Carviz s.n.* (RB); Tijuca, *Emygdio & Smith 3086* (R); estrada para Friburgo, Km 60, *Emygdio 1134* (R, UC); Guanabara, estrada dos Bandeirantes, *Ferreira 94* (GUA); Jose Bulões, *Fromm & Santos 1278* (R); circa Rio de Janeiro, *Gaudichaud s.n.* (K); estrada Rio–Petropolis, Km 37, *Luz 1563* (R, RB); Rio de Janeiro, *Sello 47* (BM); Serra Mararcu, *Schott 4148* (W); Magé, antiga Fazenda da fabrica de papel, (Paz. Pedro Oca), proxima ao Rio Gananal, ca. 23°S, 40°′W, *Souza et al. 628* (GUA, US); Serra dos Õrgaos, *Ule 2394* (R); Gavêa, *Ule s.n.* (R); Fazenda da Mocinha, *Vianna Freire 789* (R).—SÃO PAULO: Morro das Pedras, Mpio. Iguapé, *Brade 8166* (R); Serra do Mar, above Ubatuba, *Davis et al. 59777* (K, MBM); prox. Caraguatatuba, São Paulo, *Hoehne 15010* (SPF); Ubatuba, trilha da colonia de ferias do DER, estrada Taubate–Ubatuba, 23°23 S, 45°7 ′W, *Leitão et al. 34632* (SPF).—Without locality: "Brasilia," *Mikan s.n.* (BR).

Aciotis paludosa is easily distinguished by the dense and long, reddish pubescence on the petioles and stems, the membranous leaves, and the caducous fruits arranged in double biparous cymes. The Brazilian collections *Pizziolo 212* and *234* (US) are labeled as having dark purple flowers.

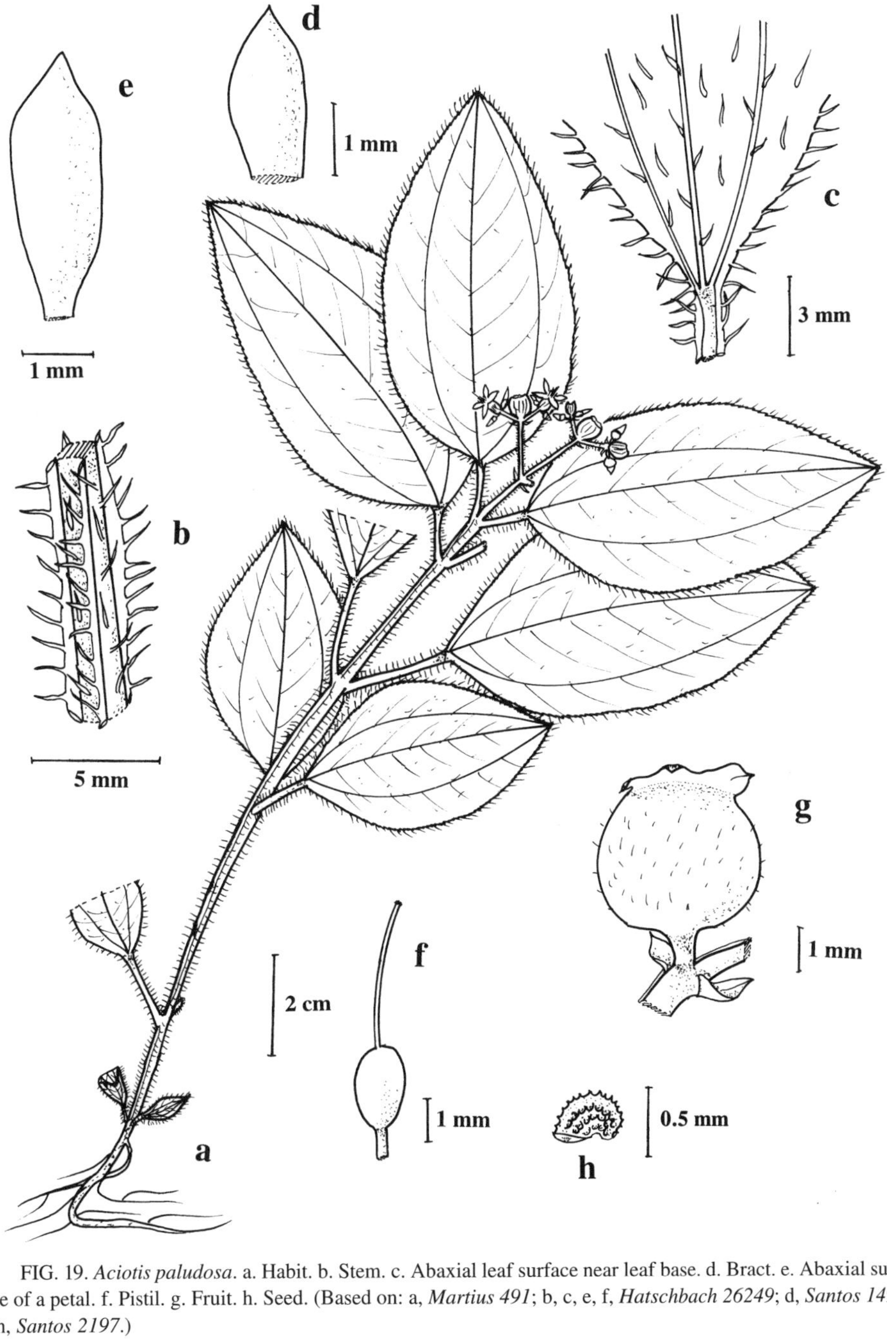

FIG. 19. *Aciotis paludosa*. a. Habit. b. Stem. c. Abaxial leaf surface near leaf base. d. Bract. e. Abaxial surface of a petal. f. Pistil. g. Fruit. h. Seed. (Based on: a, *Martius 491*; b, c, e, f, *Hatschbach 26249*; d, *Santos 149*; g, h, *Santos 2197*.)

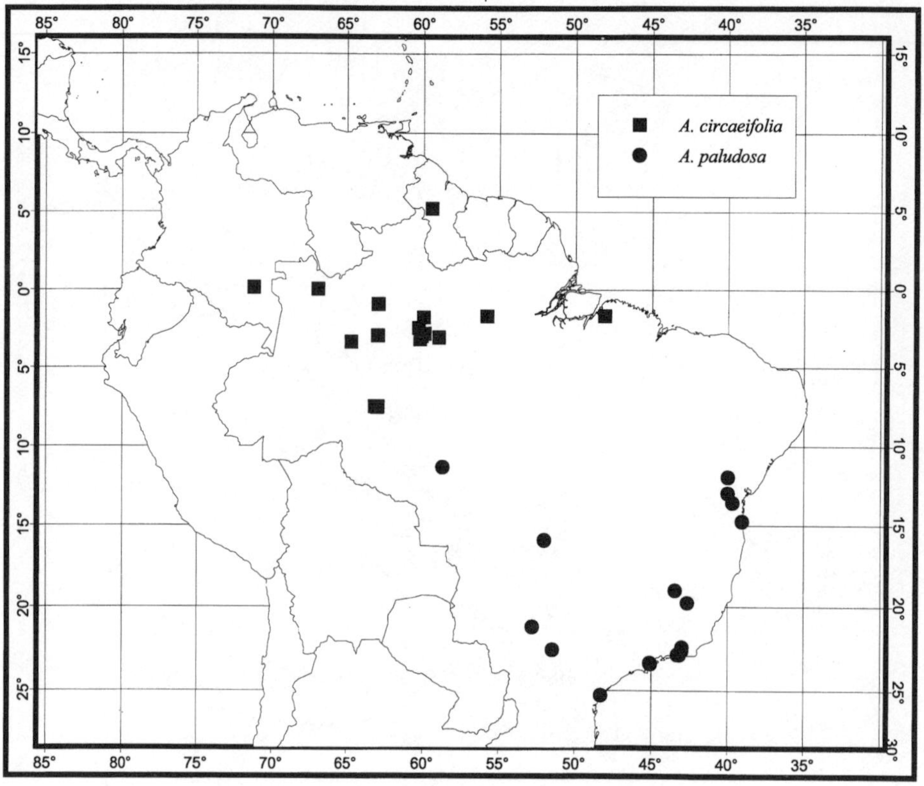

FIG. 20. Distribution of *Aciotis circaeifolia* and *A. paludosa*.

Although most of the collections have relatively small leaves, a Brazilian collection from Bahia, *Santos 2197* (US), has larger leaves (14 × 5 cm) and a laxer inflorescence with up to three pairs of first order paraclades. Wurdack identified this collection as *A. purpurascens*, and Mathies (1981) thought it belonged to an undescribed species. This specimen is *A. paludosa*; perhaps the leaves are so large, because the plant was growing in the shade in a cocoa plantation. A specimen of *A. purpurascens* that was growing in the primary forest in the Jatun Sacha Biological Station (Napo, Ecuador) also had leaves much larger and thinner than those of individuals growing in sunnier and more open areas in the same locality.

An illustration of *A. paludosa* is found in Raddi (1828; as *Rhexia herbacea*), which shows a biparous cyme, a character rather uncommon in the species. The original drawing of this plate is mounted with some specimens of *A. paludosa* on a sheet labeled *Mikan s.n.* (BR). Mikan and Raddi participated in the Austrian expedition to Brazil between 1817 and 1819 (Stafleu & Cowan 1981), so it is possible that this Mikan collection is a paratype.

Candolle cited two syntypes for *Spennera brachybotrya*, *Wied-Neuwied s.n.* (in the Kunth Herbarium, formerly at B, now destroyed) and *Martius s.n.* (G-DC). The specimen at G-DC (microfiche!) is here selected as the lectotype; a specimen deposited at BR is an isolectotype.

Naudin (1850) did not cite a type in the protologue for *Spennera ferruginea*; however, photographs of *Bonpland 49* (P) show a label with the description and drawings of the flower, both in Naudin's hand. This specimen is considered the holotype.

In the protologue of *A. spiritusanctensis*, Brade cites *Mendonça & Vieira 28* as the type collection, but the holotype is labeled "*J. Vieira & J. Mendonça 28*."

7. Aciotis acuminifolia (Martius ex DC.) Triana, Trans. Linn. Soc. London 28: 51, pl. 3, fig. 36c. 1871. *Spennera acuminifolia* Martius ex DC., Prodr. 3: 116. 1828.—TYPE: BRAZIL. "in udis Brasiliae aequatorialis," *Martius s.n.* (holotype: M!, photos: F! G! GH! MO! NY!).

Spennera dichotoma Bentham, J. Bot. (Hooker) 2: 295. 1840. *Aciotis dichotoma* (Bentham) Cogniaux in Eichler, Fl. bras. 14(3): 460, pl. 104, fig. 1. 1885.—TYPE: GUYANA. "swampy situations on the Essequibo and Rupunoony" [= Rupununi], May 1838, *R. H. Schomburgk I 513* (holotype: K!; isotypes: BM! BR! E! F! G! U, W!, digital image of U isotype: MO!).

Spennera dichotoma var. *lanceolata* Naudin, Ann. Sci. Nat. Bot., sér. 3, 14: 142. 1850.—TYPE: SURINAME. "in uliginosis Guyanae batavicae prope Oude Reiding," *Kegel Cat. no. 387* (holotype: P).

Spennera anomala Miquel, Linnaea 18: 274. 1844. *Aciotis dichotoma* var. *anomala* (Miquel) Cogniaux in Eichler, Fl. bras. 14(3): 461. 1885.—TYPE: SURINAME. "ad Osembo [ca. 05°35′N, 55°10′W]," Apr 1842, *Focke 584* (holotype: U, digital image: MO!; isotype: K!).

Spennera sphaeranthera Naudin, Ann. Sci. Nat. Bot., sèr. 3, 14: 142, pl. 5, fig. 2. 1850. *Aciotis sphaeranthera* (Naudin) Cogniaux in Eichler, Fl. bras. 14(3): 458. 1885.—TYPE: PERU. Cuzco: "prope Cuzco" [13°31′S, 71°58′W], 1839, *Gay s.n.* (holotype: P, photos: F! NY!, fragment: F!).

Aciotis aequatorialis Cogniaux in Eichler, Fl. bras. 14(3): 464, pl. 105, fig. 2. 1885.—TYPE: BRAZIL. Amazonas: "in vicinibus Barra, Prov. Rio Negro", Jan 1851, *Spruce 1152* (lectotype, here designated: M!).

Aciotis amazonica Cogniaux in Eichler, Fl. bras. 14(3): 461, pl. 105, fig. 1. 1885.—TYPE: BRAZIL. Amazonas: Ega [=Tefé, 03°24′S, 64°45′W], 1831, *Poeppig 2919* (holotype: BR!, fragment: BR!; isotypes: B, destroyed, G! NY! W!, photos of B isotype: F! GH!, fragment of B isotype: F!).

Aciotis amazonica var. *radicans* Cogniaux in Eichler, Fl. bras. 14(3): 462. 1885.—TYPE: BRAZIL. Amazonas: "prope Panuré ad Rio Uaupés" [00°30′N, 67°03′W], Sep 1952, *Spruce 2451* (holotype: BR!; isotypes: K! NY! W!, photos of W isotype: F! NY!).

Aciotis dichotoma var. *longifolia* S. Moore, Trans. Linn. Soc. London, Bot. 4: 361. 1895.—TYPE: BRAZIL. Mato Grosso: Sta. Cruz [16°00′S, 56°00′W], Dec 1891, *Moore 730* (holotype: BM!; isotypes: B destroyed, K! NY! R!, photos of B isotype: F! GH! NY!).

Herbs (2–) 3–30 (–50) cm tall or long, erect to prostrate; stolons present or absent; young stems quadrangular with inconspicuous wings, pubescent with whitish to pinkish eglandular (glandular) trichomes <1 mm long, flattened at the base. Leaf blades 1–6.5 (–9.5) cm long, 0.5–3.5 (–4.5) cm wide, angular-ovate to elliptic, apex acute to acuminate, base truncate to attenuate-decurrent, margin serrulate, membranous, 3 (–5)-veined; glabrous or pilose when young; when mature adaxially dark green, glabrous or with whitish eglandular trichomes, abaxially green, glabrous or with whitish eglandular trichomes, without web-like striations; petiole 0.8–3.5 cm long, with narrow wings, glabrous or with whitish to pinkish eglandular trichomes <1 mm long. Inflorescence a cyme or a double biparous cyme, ca. 6 cm long, 3.5–6 cm wide, with 1–2 (–3) pairs of first order paraclades; rachis glabrous or with glandular trichomes; internodes 1–1.8 cm long; first order hypopodia 1–2.2 cm long, biparous; second order hypopodia 0.5–1 cm long, biparous or uniparous; third order hypopodia 0.5–0.8 cm long, uniparous (rarely biparous); fourth order hypopodia absent to ca. 0.5 cm long, uniparous (rarely biparous); terminal paraclade 0.8–2 cm long, with 1–2 flowers per cm; bracts ca. 1.5 mm long, 0.6–1 mm wide, ovate, apex acute, abaxially glabrous or with a few glandular trichomes. Hypanthium 2.5–3.5 mm long, 1.6–2 mm in diameter, angled, without filiform sclereids in parenchyma, abaxially with glandular trichomes on the costae, adaxially glabrous. Petals 2–3.5 (–4.5) mm long, 0.8–1.8 mm wide, obovate, apex obtuse to acute, base attenuate, with or without a terminal glandular trichome, abaxially glabrous. Stamens with filaments 2.2–3.5 mm long, anthers 0.5–1.2 mm long, pinkish, pedoconnective absent to ca. 0.5 mm long. Ovary without filiform sclereids in parenchyma, apex glabrous. Fruit a fleshy capsule, pericarp caducous, placenta persistent after dehiscence, 3–4 mm long, 2.5–3.2 mm in diameter, bilocular, pericarp membranous, green when ripe, translucent; seeds ca. 0.40 mm long, 0.20–0.30 mm wide. Figs. 3g, h, 4d, 5e–h, 6d, 21.

Phenology. Flowering and fruiting throughout the year.

Distribution (Fig. 22). Lesser Antilles and South America (except Chile, Argentina, Paraguay, and Uruguay); disturbed dry or swampy areas; from sea level to 1000 m.

Vernacular names. *Charapa sacha*, *muchaquilla*, *rifarillo*, *shashak* (Peru: Loreto); *chulco* (Ecuador: Pichincha).

ADDITIONAL SPECIMENS EXAMINED. **Martinique**. Ajoupe–Bouillon, 14°30′N, 60°55′W, *Duss 4638* (NY, BR); Ajoupa-Bouillon, *Kimber 723* (WIS). **Grenada**. Grand Etang, 12°05′N, 61°42′W, *Howard 10646* (BM, GH); St. Andrew Parish, *Proctor 17029* (BM, IJ, US). **Trinidad**. Arena Forest Reserve, *Adams 14204* (TRIN); Cascada St. Ann's, 10°10′N, 62°00′W, *Broadway 5088* (BM, F, G, MO).

Colombia. AMAZONAS: road to Tarapacá, Km 17, *Lleras 458* (COL); Leticia, Corregimiento de Tarapacá, Parque Nacional Natural Amacayacu, *Rudas et al. 2801* (MO, US); Río Popeyaca (tributary of Apaporis between Piraparana and Raudal Yayacopi, *Schultes & Cabrera 15609* (NY, US).—BOYACÁ: Pajarito, 05°15′N, 73°30′W, *Uribe 4533* (COL, NY).—CHOCÓ: Hoya del Río San Juan, Road Andagoya–Condoto, 05°06′N, 76°40′W, *Forero et al. 5244* (COL); Quibdó–Guayabul road 3–6 km N of Quibdó, *Gentry & Rentería 24159* (COL, MO, US).—GUAVIARE: Mpio. El Retorno, Cerritos–El Retorno, 02°20′N, 72°41′W, *Cárdenas et al. 6707* (COAH, MO).—META: Intendencia El Meta, ca. 20 km SE of Villavicencio, 04°04′N, 73°30′W, *Killip 34255* (BM, US).—NORTE DE SANTANDER: Cordillera Oriental, Sarare, between El Caraño and El Indio, 07°10′N, 72°10′W, *Cuatrecasas 13012* (US).—VAUPÉS: Guaviare, Mpio. San José del Guaviare, Inspección de La Carpa, vd. Las Orquídeas, 02°25′N, 72°55′W, *López & Giraldo-C. 690* (MO).—VICHADA: Quinche, Caquetá River, *Edén 44* (K); Comisaría del Vichada, San José de Ocune, *Hermann 11046* (COL, NY); Parque Nacional Natural El Tuparro, border of Caño Hormiga ca. 6 km N of El Tapón, 05°11′N, 96°09′W, *Zarucchi & Barbosa 3812* (MO, US). **Ecuador**. COTOPAXI: Quevedo–Latacunga road, Km 46 from Quevedo, 00°55′S, 79°11′W, *Holm-Nielsen et al. 2855* (AAU, F, GB, MO, NY, S, US).—EL ORO: La Maná–Guayacán–Latacunga road, *Freire-Fierro 2775* (MO, QCA).—LOS RÍOS: Playa de Piedra, *Dodson 5692* (MO); Quevedo, Río Palenque Biological Station, 00°35′S, 79°22′W, *Dodson 6283* (RPSC, SEL).—NAPO: Tena, *Asplund 8848* (MO, S); Archidona–Sarayacu, *Freire-Fierro & León 2759* (QCA); Jatun Sacha, 01°04′S, 77°36′W, *Freire-Fierro 2842* (MO, QCA); Nalpi–Canelos, at Río Bobonaza, NW of Canelos, *Lugo 1484* (A, F, GB, WIS, US, NY, UC).—PASTAZA: Shell–Madretierra–La Isla–Palora rd, Km 10, *Freire-Fierro 2787b* (MO, QCA); Shell–Mera rd, Km 8, Río Alpayacu, *Freire-Fierro 2787* (QCA); Puyo–Veracruz, detour to El Calvario and Las Antenas, *Freire-Fierro 2797* (MO, QCA), 2803 (QCA); Puyo–Macas rd, detour to Puyopungo, *Freire-Fierro 2805* (MO, QCA); Madretierra–La Isla, *Freire-Fierro 2989* (QCA); Mera, *Asplund 18663* (S), *Harling et al. 19614a* (GB); El Puyo, 01°28′S, 77°59′W, *Hudson 780* (MO); Montalvo, *Løjtnant & Molau 13318* (AAU); El Puyo, *Lewis et al. 13509* (MO, QCA); Nalpi, Bobonaza river, between Cabeceras and Canelos, 01°38′S, 77°36′W, *Lugo 1467* (S, US);

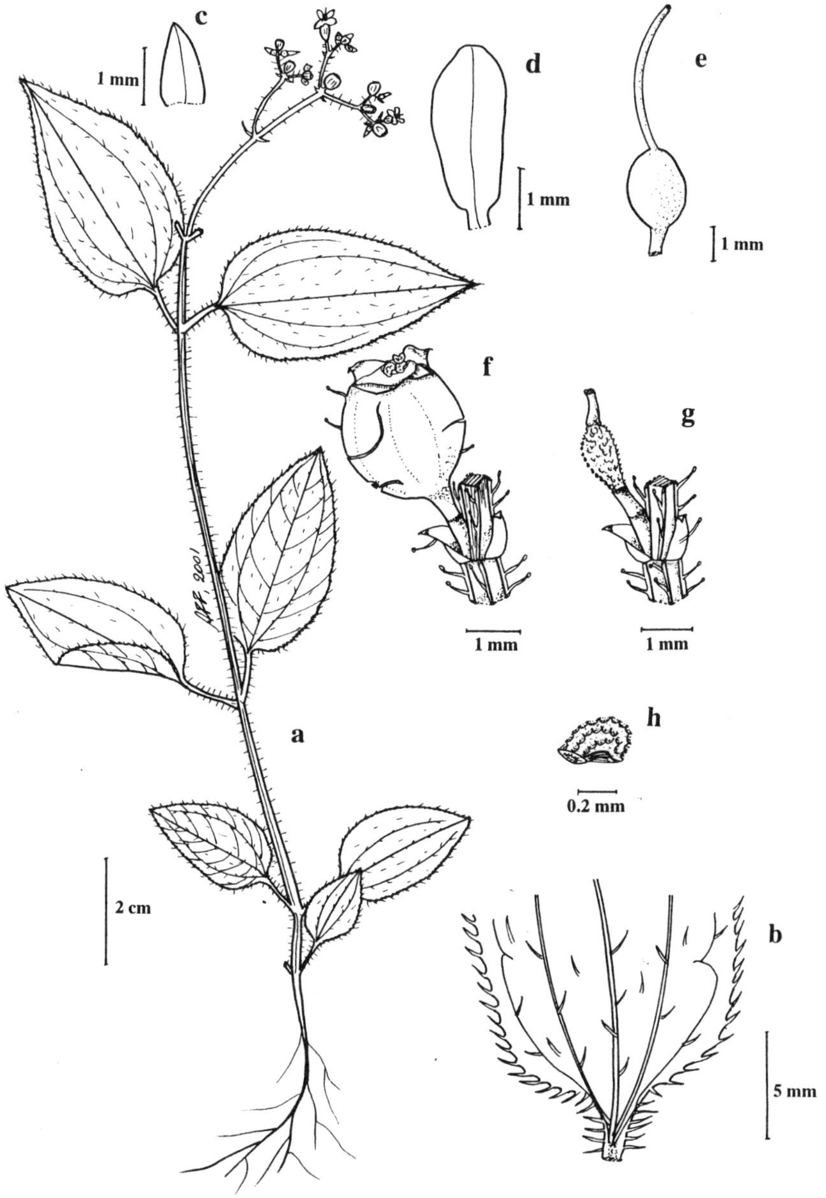

FIG. 21. *Aciotis acuminifolia*. a. Habit. b. Abaxial leaf surface near leaf base. c. Bract. d. Abaxial surface of a petal. e. Pistil. f. Fruit. g. Persistent placenta after pericarp and seeds had fallen off. h. Seed. (Based on: a, *del Carpio 1704*; b, h, *Asplund 8848*; c, d, f, *Irwin et al. 21228*; e, *Gentry & Berry 14639*; g, *Metcalf 30664*.)

Puerto Ubilla, vicinity of Pacayacu, *Lugo 1696* (AAU, F, GB, MO, US); El Puyo, *Skutch 4529* (A, F, K, MO, NY, US). **Peru**. LORETO: Maynas, *Ayala 2707a* (F, MO, NY); Distrito Nauta, Reserva Nacional Pacaya-Samiria, *del Carpio et al. 1704* (MO); Puritana, *Fosberg 29067* (US); Mishuyacu, *Killip & Smith 26888* (F, NY, RB, US); Iquitos, 03°50′S, 73°15′W, *Killip & Smith 29852* (MA, NY, S, US); Mishuyacu, *King 1188* (F, NY); Río Corrientes, 1 km S of junction with Río Macusari, *Lewis et al. 9990* (US); Requena, *Meza 505* (AAU); "ad scaturigines pagi Loreto in ripis flum.," *Poeppig s.n.* (G); Madre de Dios Explorer's Inn, near confluence of rivers Tambopata and La Torre, 12°50′S, 79°20′W, *Smith et al. 1103* (US); Bora Native community of Brillo

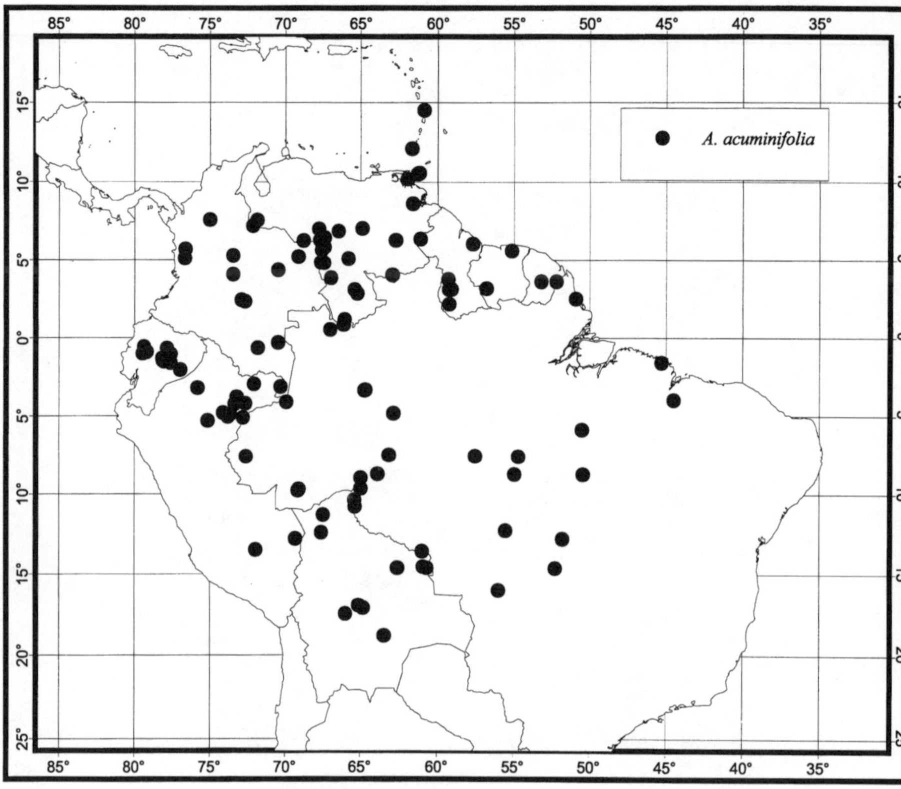

FIG. 22. Distribution of *Aciotis acuminifolia*

Nuevo, Yaguasyacu River, *Treacy & Alcorn 416* (F, WIS); Puesto Hamburgo (Río Samiria), *Vásquez et al. 6391* (MO); Requena, road Requena–Yaquerana, *Vásquez 13377* (MO); Mishuyacu, *Williams 1486* (F, NY). **Bolivia.** COCHABAMBA: Cantón Ivirgarzama, *Beck 13728* (US); José Carrasco Torrico, valle del Sajta, 17°08′S, 64°50′W, *Kessler et al. 8830* (MO); Chapare, Chimoré, *Steinbach 451* (F, GH, MICH, MO, NY, US).—EL BENI: 11–15 km NW of Guayaramirím on road to Cachuela Esperanza, 10°48′S, 65°23′W, *Anderson 12044* (MO, NY, US); *Meyer 255* (Z).—LA PAZ: Abel Iturralde, Siete Cielos, Río Manupare, 12°27′S, 67°37′W, *Solomon 16927a* (MO).—PANDO: Manuripi, 11°20′S, 67°30′W, *Beck et al. 19416* (US).—SANTA CRUZ: Prov. Ibañez, ca. 22 km towards Abapó, *Beck 6527* (US); Serranía de Huanchaca, 14°33′N, 60°56′W, *Foster et al. 13832* (LPB); Prov. Velasco, estancia Flor de Oro, *Nee 41297* (NY, US); Prov. Ñuflo de Chávez, 5 km N Perseverancia, *Vargas I. 564* (MO, NY). **Venezuela.** AMAZONAS: Manapiare, *Agostini 1519* (VEN); Río Mawarinuma, *Anderson 13330* (CAS, F, MICH); Río Mawarinuma, 00°50′N, 66°10′W, *Boom & Weitzman 5933* (AAU, GH, MO, NY, VEN); Pto. Ayacucho–Sanariapo, Km 35, *Bunting et al. 3460* (US); 9–10 km from Pto. Ayacucho, *Castillo 776* (MO, US, VEN); Río Mawarinuma, *Davidse & Miller 26929* (MICH, MO, NY, US); Río Negro, between Neblina Base Camp and the mouth of Canyon Grande at Puerto Chimo along the Río Mawarinuma, *Davidse & Miller 27048* (MO, US); 0–3 km from Río Sipapo, *Gentry & Berry 14639* (MO, US, VEN); Pto. Ayacucho–El Burrol, Km 28, *Gröger & Llamozas 1124* (US); Atabapo, Río Ocamo, 02°50′N, 65°14′W, *Guanchez 1957* (MO); Depto. Atures, Ceguera, *Guanchez & Melgueiro 3641* (MO, US); Orinoco, between San Fernando de Atabapo y San Antonio, *Level 52* (US); Río Yatua above mouth of Yacibo, *Maguire et al. 36539* (F, US); Río Mawarinuma, *Renner 2105* (VEN); Pto. Ayacucho–Samariapo, *Stergios 3215* (US); Táchira, Cerro La Espuma, of La Espuma, SW of Sto. Domingo, *Steyermark & Liesner 119219* (MO, US, VEN); south-facing, shaded sandstone bluffs along old highway, just W of Ayarí, *Steyermark et al. 119490* (MO, US, VEN); Río Mawarinuma, *Steyermark & Luteyn 129721* (MO, US).—APURE: Distrito Pedro Camejo, along banks of Río Capanaparo, *Davidse & González 13097* (MO); Distrito Pedro Camejo, near Río Meta, Morichal de la Madera, 06°13′N, 68°49′W,

Davidse & González 14201 (MO, VEN).—BOLÍVAR: Depto. Cedeño, Km 12–120 S from road Caicara del Orinoco–Pto. Ayacucho, *Aymard & Stergios 3326* (AAU, MO); Mpio. Sucre, Río Caura, Jabillal, *Fernández & Sanoja 5111* (US); Reserva Forestal Imataca, entre Isla Anacoco y Araguay, 06°20′N, 61°09′W, *Stergios et al. 2737* (US); Depto. Cedeño, Caño Garrapata, *Stergios et al. 7228* (MO, US); Salto de Chalimano, Río Paramichi, 18 km N from Venezuela-Brazil border, *Steyermark 90719* (F, NY, US, VEN); Distrito Piar, Laguna de Canaima, *Steyermark et al. 127966* (US, VEN).—DELTA AMACURO: Sacupana, 08°35′N, 61°39′W, *Rusby & Squires 40* (BM, E, F, G, GH, K, M, MO, NY, Z). **Guyana**. SE Kanuku Mts, foothills S of Crabwood Creek, ± 2. 5 km E of Makawatta Mtn, *Hoffman 385* (NY, US); Takutu-Essequibo, Northern Rupununi, *Hoffman & Capellaro 1009* (NY, US); Rupununi, Kanuku Mts., 03°07′N, 59°06′W, *Jansen-Jacobs et al. 3388* (K); Kuyuwini Landing, Kuyuwini River, *Jansen-Jacobs et al. 2481* (NY, US); Berbice River, *Maas et al. 5453* (US). **Suriname**. Wilhelmina Gebergte, Zuid River, Kayser Airstrip 45 km above confluence with Lucie Rivier, 03°10′N, 56°49′W, *Irwin et al. 55256* (NY, US). **French Guiana**. Saül, *Gracie 19019* (US); route de Bélizon, 0–2 km S of Eaux Claires, *Mori et al. 22043* (CAY); near Eaux Claires, 03°37′N, 51°12′W, *Mori et al. 22332* (CAY, US). **Brazil**. ACRE: Igarapé Preto, ± 5 km from airport, *Araújo & Ferreira 440* (HRB, MG); Cruzeiro do Sul, along Moa, 07°38′S, 73°36′W, *Croat & Rosas 62436* (INPA); Mpio. Sena Madureira, Rio Macauã, Seringal Riozinho, *Lima et al. 567* (NY, US).—AMAPÁ: Rio Calcoene, Sete Ilhas, 02°30′N, 50°55′W, *Pires & Cavalcante 52582* (K, NY, US).—AMAZONAS: Beaba, Rio Purus, 04°52′S, 62°53′W, *Gottsberger & Döring 19-12286* (MO); in vicinia Manaos, *Spruce 1191* (K).—MARANHÃO: São Mateus do Maranhão, ca. 42 km N from Peritoró, *Jangoux & Bahia 1098* (US); Nova Esperança, Rio Alto Turiaçu, *Jangoux & Bahia 690* (MBM, MG); Mpio. Humaitá, *Janssen 248* (M); Rio Solimões and Rio Javari, *Lleras et al. P16895* (AAU, F, GH, K, MG, MO, NY, S, US); Mpio. Tefé, 03°22′S, 64°42′W, *Plowman et al. 12071* (INPA, NY); Serra dos Carajás, *Sperling et al. 6172* (F, MG, NY, US).—MATO GROSSO: Córrego do Gato, *Harley et al. 10445* (K, MO, NY, R, UC, US); Xavantina, N side of Rio das Mortes, 14°38′S, 52°14′W, *Philcox & Freeman 4741* (K, US); Mpio. Sinop, 7 km E of BR-163 on road N of Rio Celeste, *Thomas et al. 3931* (F, INPA, NY, SPF, US).—PARÁ: SE of Missão Cururú, *Anderson 10621* (C, F, K, MO, NY, US); Mpio. Sta. Ana do Araguaia, 100 km S of Redenção on road (PA-150) to Barreiras dos Campos, 08°45′S, 50°25′W, *Plowman et al. 8917* (F, GH, HRB, INPA, MICH, MO, NY, US); estrada Santarém/Cuiabá, Km 877 de Cuiabá, Cachoeira da Luz do Rio Curua, *Silva 201* (INPA).—RONDÔNIA: Porto Velho, *Nee 34879* (US); Mutumparana–Rio Madeira, *Prance et al. 5506* (F, INPA, K, MG, NY, R, S); 10 km de Porto Velho, 08°45′S, 63°54′W, *Rodrigues 10443* (INPA, UEC).

Aciotis acuminifolia, an easily recognized species, is characterized by the eglandular pubescence on the stems, the 3-veined membranous leaves, which are usually glabrous and ovate, the simple or double biparous cymes with up to three pairs of paraclades, the fleshiness of the hypanthium and persistence of the placenta, and the small ovoid anthers with pedoconnectives half as long as the anther. It is a widely distributed species that shows some variability in stature, inflorescence size, leaf shape, and pubescence density. Densely pubescent plants with somewhat lax inflorescences could be confused with *A. paludosa* or *A. ornata*. Yet, in *A. paludosa* the leaf apex is never acuminate, and pericarps and placentas are caducous; *A. ornata* has the leaf apex mostly obtuse, and the mature fruits are never translucent when dry. Glabrous specimens of *A. purpurascens* with attenuate leaves and greenish, translucent fruits could also be mistaken for *A. acuminifolia*; however, *A. purpurascens* is taller (at least 50 cm), and has thyrsoid inflorescences and caducous fruits.

Aciotis acuminifolia may hybridize with *A. polystachya*; see Appendix 2.

The morphological variation of *A. acuminifolia* is reflected by the many names, here placed in synonymy, that have been applied to this taxon. The types of *A. acuminifolia*, *A. amazonica*, and *A. dichotoma* are glabrous plants with elliptic leaves and double biparous cymes; the type collection of *A. dichotoma* var. *anomala* comprises glabrous plants with elliptic leaves and cymose inflorescences. On the other hand, the type of *A. sphaeranthera* is pubescent, and has elliptic leaves and cymose or double-cymose inflorescences. This morphological plasticity perhaps is caused by differences in soils and humidity. Populations from very humid and shaded areas consist of large and glabrescent plants with inflorescences having up to three pairs of paraclades (e.g., *Freire-Fierro 2775*), whereas populations from open and sunnier areas include pubescent individuals as small as 2 cm tall and with only one to three terminal flowers (e.g., *Freire-Fierro & León 2759*).

The name *A. amazonica* has been applied widely to glabrous specimens from Colombia and Venezuela. Two Venezuelan collections from Laguna Canaima (*Hertel & Oberwinkler 15217*, *Steyermark et al. 127966*) differ in having narrowly elliptic leaves, but otherwise conform to *A. acuminifolia*. An exceptional collection from Colombia, *Idrobo 492*, has cordate leaves up to 9.2 cm long.

Macbride (1941), in his treatment of Melastomataceae for the *Flora of Peru*, listed six collections under "*A. cordata*" (see Doubtful and Excluded Names). Of these five belong to *A. acuminifolia* (*Killip & Smith 26888*, *Klug 1188*, *Osgood 16*, *Williams 1710*, *1712*); the remaining one, *Schunke 213*, is *A. rubricaulis.*

In the protologue of *A. aequatorialis* Cogniaux listed the Brazilian collections *Spruce 816*, *Martius s.n.*, *Spruce 1191*, and *Spruce 1152*, as well as the Peruvian collection *Poeppig s.n.* Of these, *Spruce 816* belongs to *Aciotis annua*; *Spruce 1152* (M) is here designated the lectotype. These syntype collections consist of glabrous or pubescent individuals with ovate leaves, but with the leaf base decurrent to cuneate. The inflorescences vary from cymes to double biparous cymes.

Although I did not see the type of *Spennera dichotoma* var. *lanceolata* I believe it is a synonym of *A. acuminifolia*. Naudin (1850) noted that the type specimen differs from the other representatives of *Spennera dichotoma* [=*A. acuminifolia*] only in its oblong-ovate leaves with an attenuate leaf base.

Good illustrations of *A. acuminifolia* may be found in Cogniaux's treatment of *Aciotis* for *Flora brasiliensis* (1885); see plates 104 and 105 (as *A. aequatorialis*, *A. amazonica*, *A. dichotoma*, and *A. martiana*).

Aciotis acuminifolia is represented by *Irwin et al. 21228* (TEX) in Whiffin and Tomb's (1972) study of seeds of the Neotropical capsular-fruited species of Melastomataceae.

Renner (1984) reports that Brazilian populations of *A. acuminifolia* are pollinated by the bees *Megommation ogilvei*, *Paratetrapedia duckei*, *Melipona fulva*, by a species of *Melipona* Illiger 1806, and by a species of the tribe Exomalopsini (Apidae:Apinae).

In Peru, "*shashak*" (*Lewis 9990*) is used for candidiasis.

8. Aciotis olivieriana Freire-Fierro, Novon 11: 168. 2001.—Type: Colombia. Meta: Villavicencio [04°09′N, 73° 37′ W], 500 m, 19 Mar 1939, *Alston 7650* (holotype: US!; isotypes: BM! COL! NY! S!).

Herbs 14–30 cm tall, erect to prostrate; stolons present or absent; young stems quadrangular with inconspicuous wings, glandular-pubescent with trichomes <0.5 mm long. Leaf blades 1.5–3 (–5) cm long, (1–) 2.3–2.8 (–3.6) cm wide, cordiform, apex acute to acuminate, base cordate, margin serrulate, membranous, 7-veined; adaxially dark green, pubescent with whitish eglandular appressed trichomes, abaxially light green, pubescent, with eglandular appressed trichomes, without web-like striations; petiole (0.3–) 1.3–3.2 cm long, with narrow wings, glandular-pubescent with trichomes <1 mm long. Inflorescence a double biparous cyme (rarely a thyrse), 6–14 cm long, (3–) 4–5.5 (–7) cm wide, with 2–3 pairs of first order paraclades; rachis glandular-pubescent; internodes 1.5–3 cm long; first order hypopodia 1–1.5 cm long, biparous; second order hypopodia 0.3–0.7 cm long, biparous or uniparous; third order hypopodia 0.2–0.5 cm long, uniparous (rarely

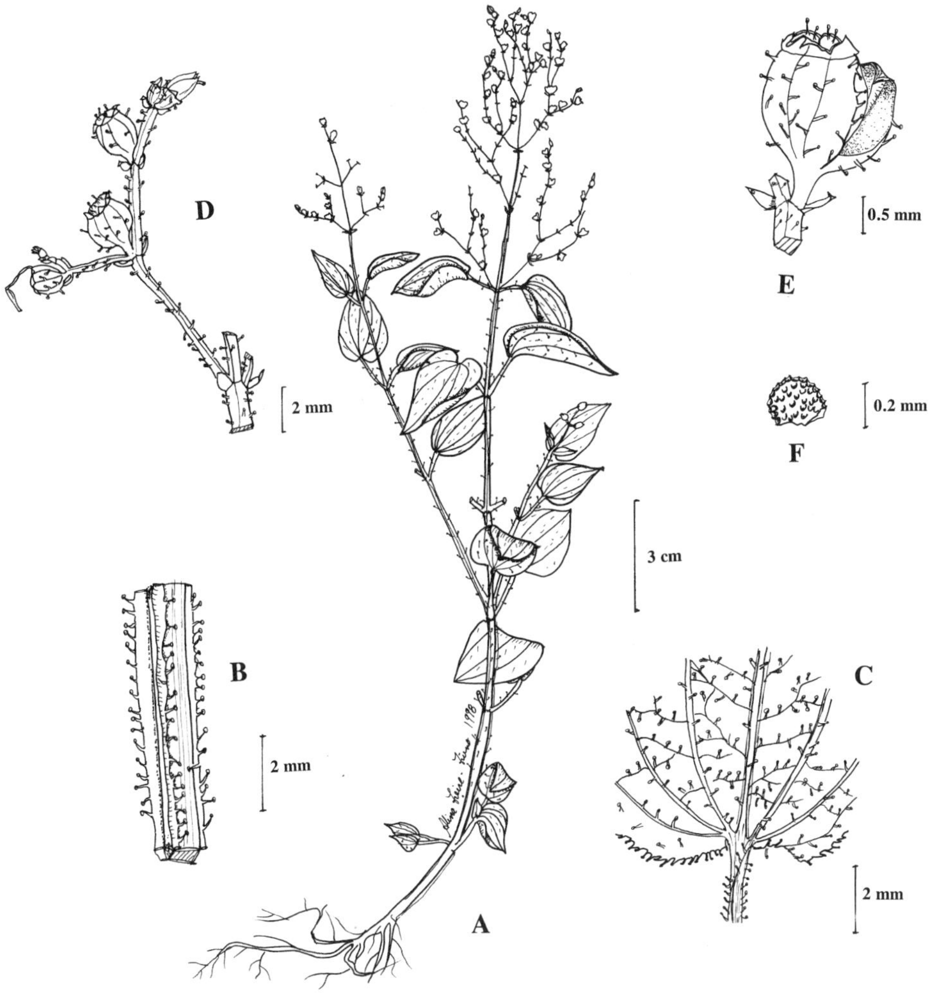

FIG. 23. *Aciotis olivieriana*. A. Habit. B. Stem. C. Abaxial leaf surface near leaf base. D. Portion of inflorescence. E. Open capsular fruit. F. Seed with foveolate testa. (Based on *Sastre 968*; reprinted from *Novon* with the permission of the Missouri Botanical Garden.)

biparous); fourth order hypopodia absent; terminal paraclade 1–2.5 cm long, with 2–3 flowers per cm; bracts 0.8–1.5 mm long, 0.2–0.5 mm wide, triangular, apex acuminate, abaxially glabrous. Hypanthium ca. 1.5 mm long, 1–1.8 mm in diameter, angled, without filiform sclereids in parenchyma, abaxially with few glandular trichomes, adaxially glabrous. Petals ca. 2.5 mm long, ca. 1 mm wide, obovate, apex obtuse, base cuneate, without a terminal glandular trichome, abaxially glabrous. Stamens with filaments 2–2.5 mm long, anthers 1.5–2 mm long, purple, pedoconnective absent to ca. 0.1 mm long. Ovary without filiform sclereids in parenchyma, apex glandular-pubescent. Fruit a capsule, pericarp and placenta persistent after dehiscence, 2.2–2.5 mm long, 1.5–1.8 mm in diameter, bilocular, pericarp coriaceous, not translucent; seeds 0.30–0.40 mm long, ca. 0.25–0.30 mm wide. Fig. 23.

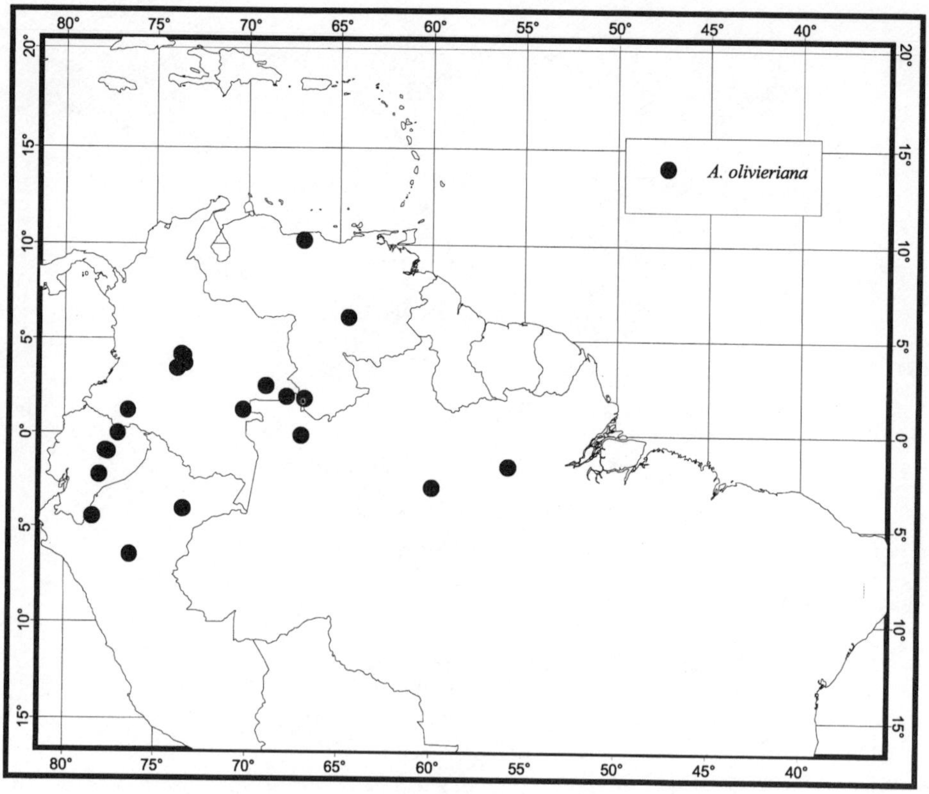

FIG. 24. Distribution of *Aciotis olivieriana*

Phenology. Flowering and fruiting from March to June and from August to January.

Distribution (Fig. 24). East of the Andes in Colombia and Ecuador, northeastern Peru, southern Venezuela, and the Brazilian Amazon Basin; riverbanks and primary forests; 125–1350 m.

Vernacular names. *Daicap* (Peru: Amazonas).

ADDITIONAL SPECIMENS EXAMINED. **Colombia**. META: Río Guatiquía, alrededores de Villavicencio, 04°09′N, 73°37′W, *Araque & Barkley 18M34* (COL, US); Villavicencio, *Cuatrecasas 1995* (NY, US), *Cuatrecasas 3505* (F); toward El Parrao, *Cuatrecasas 4584* (F, US); opposite Hda. San Carlos, 1/2 km SW Villavicencio, *Hermann 11133* (COL, NY, US); Sabanas de San Juan de Arama, margen izq. del Río Güejar, aterrizaje Los Micos, 03°25′N, 73°50′W, *Idrobo & Schultes 579* (COL); Sierra La Macarena, Bocas del Río Duda, *Idrobo 8365* (COL); El Mico Airstrip (Camp 1), *Philipson et al. 1401* (BM, COL, NY, US); Río Sardinata, *Sastre 968* (COL); Villavicencio, *Triana 6195a* (BM).—PUTUMAYO: *Croat 51747* (COL, MO).—VAUPÉS: San Felipe, Comisaría del Vaupés, 02°30′N, 69°00′W, *Romero-Castañeda 1213* (COL, NY). **Venezuela.** AMAZONAS: Mavaca, Alto Orinoco, Indios Guaicas (Yanomanö), *Aristeguieta & Lizot 7384* (NY, US); Alto Río Orinoco, between Isla Quiratare and Esmeraldas, *Maguire & Wurdack 34625* (K, NY, US); San Carlos de Río Negro, 01°56′N, 67°53′W, *Steyermark & Bunting 102703* (US); alrededores de San Carlos del Rio Negro, *Stergios & Aymard 7265* (MO, US).—BOLÍVAR: Salto Pará, 06°10′N, 64°30′W, *Meier 240* (US).—DISTRITO FEDERAL: Caracas, 10°35′N, 66°56′W, *van Landsberge 19* (S). **Ecuador**. NAPO: between Tena and Napo, *Asplund 9349* (S); Hacienda Dos Ríos, bis Río Misahuallí, Richtung Archidona, 01°02′S, 77°40′W, *Heinrichs 416* (G, M, NY, Z); about 11 km ESE of Puerto Misahuallí, ca. 3 km E of Jatun Sacha, *Renner 69334* (AAU).—ZAMORA-CHINCHIPE: Macas, 02°17′S, 78°08′W, *Asplund 19836* (S). **Peru.** AMAZONAS: Labanda, Huampami, Río

Cenepa, 04°30′S, 78°30′W, *Ancuash 696* (GH, XAL).—SAN MARTÍN: prope Tarapoto, 06°30′S, 76°25′W, 1856, *Spruce s.n.* (K). **Brazil**. AMAZONAS: Manaus, estação da Silvicultura, INPA Km 45, 02°53′S, 59°58′W, *Miranda et al. 218* (INPA); São Gabriel, upper Rio Negro, *Renner 363* (INPA); Manaus–Caracarai, Km 45, *Renner 955* (AAU); Reserva Forestal Ducke, Manaus–Itacoara, Km 26, *Sothers 417* (INPA), *Sothers 1011* (INPA); prope São Gabriel da Cachoeira [00°07′S, 67°05′W], ad Rio Negro, *Spruce 2123* (BM, K, M, NY, US); Waupes, oberer Rio Negro, *Vogel et al. 341* (INPA, US).—PARÁ: Rio Trombetas shore, Porteira cemetery, 01°45′S, 55°50′W, *Campbell et al. P22492* (K, MG, NY, S, US).

Aciotis olivieriana generally is similar to *A. polystachya*, but lacks the web-like striations on the abaxial leaf surface and filiform sclereids. It could be confused with *A. viscida*, but can be distinguished from that species by its smaller, membranous leaves, double cymose inflorescences, and absence of glandular trichomes on the abaxial surface on the petals. Similarly, *A. olivieriana* could be mistaken for cordate-leaved representatives of *A. acuminifolia*, but it differs from this species in having 7-veined leaves, glandular trichomes on the stems and petioles, and elongate anthers lacking pedoconnectives. Wurdack (1993), in his treatment of Melastomataceae for *Flora of Ecuador*, assigned *Asplund 9349* and *Asplund 19836* to *A. indecora*, but they belong to *A. olivieriana*. The two species share glandular-pubescent stems and an ovary crowned by glandular trichomes; however, *A. indecora* has petals abaxially glandular-pubescent and baccate fruits, whereas *A. olivieriana* has glabrous petals and capsular fruits.

9. Aciotis circaeifolia (Bonpland) Triana, Trans. Linn. Soc. London 28: 52. 1871. *Rhexia circaeifolia* Bonpland, Monogr. Melast. 2: 101, t. 39. 1818. *Spennera circaeifolia* (Bonpland) DC., Prodr. 3: 116. 1828.—TYPE: BRAZIL. "Brasilia," *Ferreira s.n.* (holotype: P).

Spennera longifolia Martius ex DC., Prodr. 3: 115. 1828. *Aciotis longifolia* (Martius ex DC.) Triana, Trans. Linn. Soc. London 28: 52. 1871.—TYPE: BRAZIL. Pará: "Para" [=Belém], *Martius s.n.* (holotype: M!).

Spennera circaeoides Martius & Schrank ex DC., Prodr. 3: 116. 1828.—TYPE: BRAZIL. Amazonas: "circa Ega ad flum. Amazonum" [=Tefé, 03°24′S, 64°45′W], *Martius s.n.* (holotype: M!).

Herbs to subshrubs 0.5–1.5 m tall, erect; stolons absent; young stems slightly quadrangular with inconspicuous wings, glabrescent to glabrous, initially with dark ferruginous glandular trichomes >1.5 mm long. Leaf blades 3–14.5 (–16) cm long, 1.6–4.5 (–5.8) cm wide, ovate to cordiform, apex acuminate, base cordate, margin serrulate, membranous, 5-veined; adaxially dark green, glabrous or with dark ferruginous glandular trichomes, abaxially green, glabrous or with dark ferruginous glandular trichomes, without web-like striations; petiole (0.8–) 1.5–5 cm long, with narrow wings and dark ferruginous glandular trichomes >1.5 mm long. Inflorescence a thyrse, 7–15 cm long, 5–6 cm wide, with 3–5 pairs of first order paraclades; rachis with dark ferruginous glandular trichomes; internodes 1.5–3.2 (–4.2) cm long; first order hypopodia 0.6–1.5 cm long, biparous; second order hypopodia 0.3–0.7 cm long, uniparous (rarely biparous); third order hypopodia 0.2–0.4 cm long, uniparous; fourth order hypopodia absent; terminal paraclade 1–3 (–5) cm long, with 3–4 flowers per cm; bracts 0.8–1.5 mm long, ca. 0.5 mm wide, triangular, the apex acuminate, abaxially glabrous. Hypanthium ca. 3 mm long, 1.5–2 mm in diameter, terete, without filiform sclereids in parenchyma, abaxially glabrous or with dark ferruginous glandular trichomes, adaxially glabrous. Petals 5–7.5 mm long, 2–2.5 mm wide, oblong to obovate, apex acute to acuminate-aristate, base cuneate, with or without a terminal glandular trichome, abaxially glabrous. Stamens with filaments ca. 2 mm long, anthers 1.8–2.4 mm long, white, pedoconnective 0.5–0.6 mm long. Ovary without filiform sclereids in parenchyma, apex glabrous. Fruit baccate, pericarp and placenta caducous when ripe, 4–4.5 mm long, 3–4 mm in diameter, bilocular, pericarp membranous, green when ripe, translucent; seeds ca. 0.45–0.55 mm long, 0.35–0.40 mm wide. Fig. 25.

Phenology. Flowering and fruiting throughout the year.

Distribution (Fig. 20). Amazon Basin (Colombia, Guyana, and Brazil); commonly in shaded areas; 25–620 m.

Vernacular names. *Buxixú da folha azeda, folha azeda, zinabre,* and *folha grande* (Brazil: Amazonas).

ADDITIONAL SPECIMENS EXAMINED. **Colombia.** AMAZONAS/VAUPÉS: Río Apaporis, entre Río Pacoa y Rio Kananarí, Soratama, *Schultes & Cabrera 13031* (US). **Guyana.** Demerara River, Greek Falls, *Jenman 3991* (K, NY, US); Kaeieteur Fall, *Schomburgk 863* (BM, BR, K, G, W), *Tutin 517* (BM). **Brazil.** AMAPÁ: Porto Terezinha, *Cowan 38461* (NY, R, RB, US).—AMAZONAS: IAN, Rio Negro, *Black 48-2895* (RB); Maués–Airport road, *Campbell et al. P22159* (NY, US); Rio Uatumã, Mpio. Itapiranga, margem direita do igarapé Catitu, *Cid Ferreira et al. 603* (INPA, NY); Manaus, *Chagas s.n.* (MG); along Manaus–Boa Vista road (BR-174), Km 415, *Coradin & Cordeiro 1081* (INPA, NY, US); Manaus–Caracarai, 2 km N of junction in road to Itacoatiara, 01°50′S, 60°01′W, *Croat 62215* (INPA); Praia Dorada, private beach on Rio Negro, near Manaus, *Davidson & Martinelli 9987* (INPA, NY, RB, US); Manaus, sede do INPA, *Falcão 1* (INPA); Tarumã road, 18 km N of Manaus, *Forero et al. 4631* (F, GH, MG, NY, R, S, US); B.A.M. Fz. do Franco, *Franco et al. s.n.* (INPA, NY); Cachoeira do Tarumã, *Jaccoud s.n.* (INPA); Mpio. Humaitá, 500 m ao Norte da BR-230, Km 2, 07°31′S, 63°10′W, *Janssen 117* (M); Tarumã Grande, 1 km N from the junction of Rio Negro and Igarapé Tarumã, *Keel & Guedes 316* (NY); Tarumã Grande, *Kuhlmann 18* (RB); Uypiranga, *Kuhlmann 938* (RB); Reserva Florestal Ducke, *Lisbôa 865* (INPA); Río Arara, *Loureiro et al. s.n.* (INPA); halfway between Manaus and Manacapuru, 03°15′S, 60°15′W, *Lowe 4026* (INPA, K); Cachoeirinha, *Maas & Maas 215* (INPA, RB); entre rios Arara e Cuieiras, *Madison et al. 25* (INPA); Lago do Atãana, below Barcellos, *Madison et al. 111* (INPA, US); Arquipiélago das Anavilhanas, Foz do Rio Ariau, *Medri 12* (INPA); Rio Negro, circa Coari Ega, *Martius s.n.* (M); São Gabriel, margem esquerda do Rio Negro, lago Ataana, *Nascimento 586* (MG, MBM, NY, US); boca do Rio Cuieiras, *Nelson et al. 882* (INPA, MG, NY); Ponta Negra, na margem esquerda do Rio Negro, 12 km acima de Manaus, *Pabst 9434* (HB, NY); Cachoeira do Tarumã, *Pereira 3485* (HB, RB); Manaus, *Prance et al. 11646* (F, GH, INPA, K, M, MG, NYS, R); Manaus–Itacoatiara Hwy, Km 26, Reserva Florestal Ducke, *Prance et al. 3636* (F, G, INPA, K, NY, R, S, US); Rio Urubu, 10 km above Manaus–Itacoatiara road, *Prance et al. 5130* (INPA, NY, US); Río Cuieras, 2 km from R. Cuieras at 2 km below mouth of Rio Brancinho, *Prance et al. 17992* (AAU, GH, MO, NY, S, US); Manaus, Km 9 da BR-17, *Rodrigues s.n.* (INPA); Igarapé do Buião, *Rodrigues 485* (INPA, NY); Manaus. Ig. do Buião, *Rodrigues & Coêlho 1191* (INPA); in vicinibus Barra, *Spruce 1153* (E, M); Mpio. Humaitá, estrada Humaitá–Labrea, Km 59, 6 km ao Norte, *Teixeira et al. 969* (INPA); Ilha das Araras, Rio Madeira, *Trail 250* (K).—PARÁ: Mar Marco, *Baker 79* (BR); Belém, *Black 47961* (RB); Marianas, *Ducke MG-1101* (HB); Utinga, *Guedes 2003* (G, R, RB); caminho para os Campos de Ariramba, varzea entre a Serra de Fe em Deus e Serra da Preciosa, *Martinelli 6807* (INPA, NY, US); Belém, Instituto Agronômico do Norte, *Murça & Black 1431* (US); João Coelho, Rio Caraparu, *Pereira 3341* (RB); Brevos, *Pires et al. 4867* (NY); Cachoeira Porteira, Mpio. Oriximiná, Reserva biológica do IBDF, 01°45′S, 55°50′W, *Rosario & Ubiratan 738* (MG); area de pesquisas ecológicas do Guama, Belém, I.P.E.A.N., *Schubert 2178* (US); Belém, *Silva 175* (F, K, US); Acará, Jacarequara, *Silva 512* (HB, MG); Tanaii, ad Rio Acará, *Spruce s.n.* (K, W).—RORAIMA: estrada Manaus–Caracarai, Km 329, Army post N of Wimari-Atoari Indian Reserve, Manaus-Venezuela Highway, *Steward et al. 12* (GH, INPA, MICH, MO); Serra Surucucú, *Ribeiro 15187 (814)* (US).

Aciotis circaeifolia is characterized by its hirsute pubescence, composed of dark ferruginous, glandular trichomes >1.5 mm long, the large, membranous, cordate leaves, and the narrowly winged stems. It is similar to *A. purpurascens* in its membranous leaves, thyrsoid to paniculate inflorescence, and baccate fruits; however, *A. purpurascens* has quadrangular stems with well developed wings, lacks the long (>1.5 mm) glandular hairs on the stems, has the leaf base attenuate to cuneate, and petals less than 5 mm long.

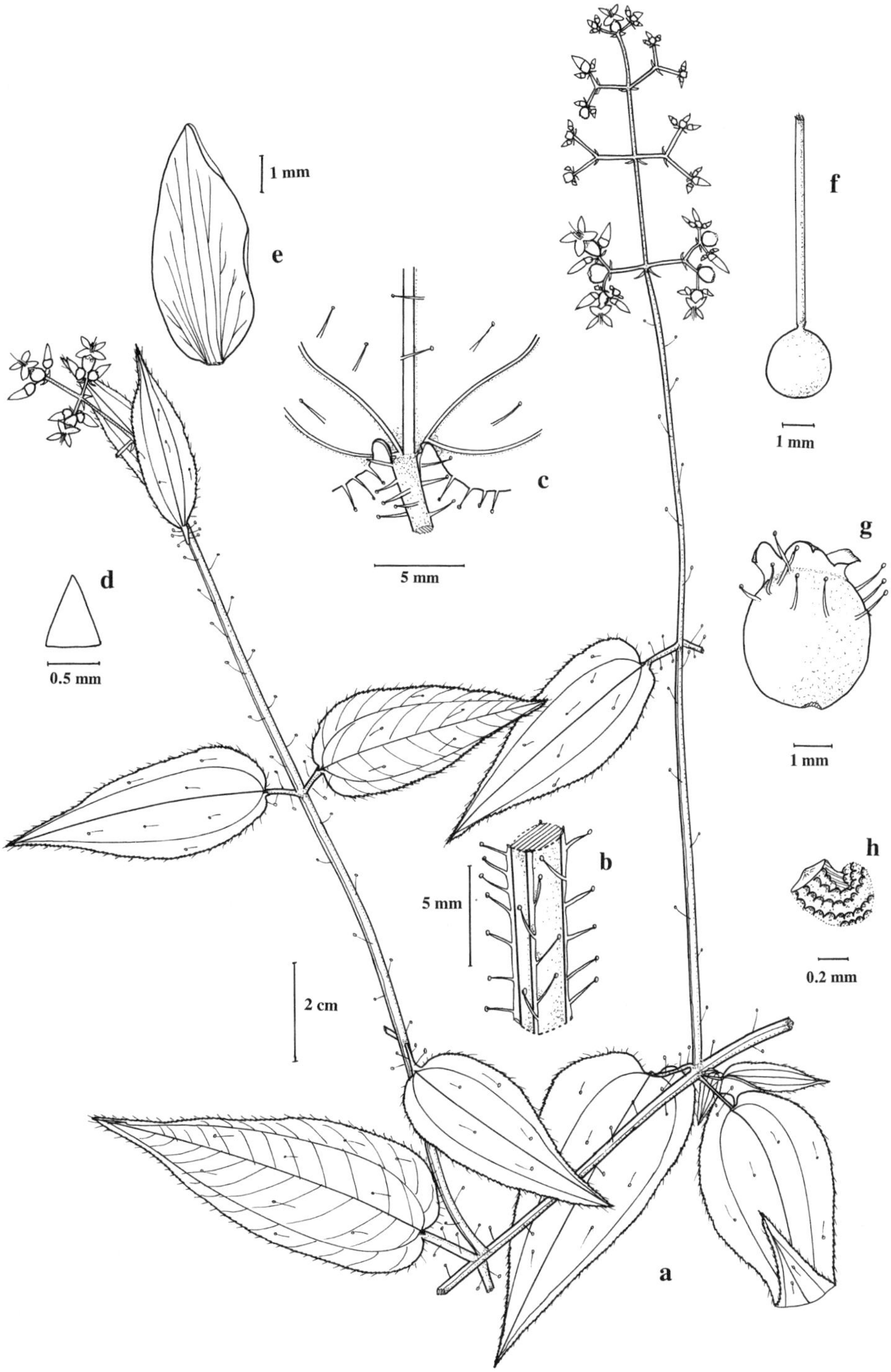

FIG. 25. *Aciotis circaeifolia*. a. Habit. b. Portion of stem. c. Abaxial leaf surface near leaf base. d. Bract. e. Adaxial surface of a petal. f. Pistil. g. Fruit. h. Seed. (Based on: a, *Steward et al. 12;* b–d, f–h, *Ribeiro & Pereira 1241*; e, *Renner 753*.)

Cogniaux (1885) provides an illustration of *Aciotis circaeifolia*; see plate 106 (as *A. longifolia*).

Renner (1984) reports that a species of *Megommation* Moure 1943 pollinates *A. circaeifolia* in Brazil.

Whiffin and Tomb (1972) used seeds of *Prance et al. 5130* (US) for their study of seeds of Neotropical capsular-fruited species of Melastomataceae.

10. Aciotis purpurascens (Aublet) Triana, Trans. Linn. Soc. London 28: 52, pl. 3, fig. 36a. 1871. *Melastoma purpurascens* Aublet, Hist. pl. Guiane 1: 402, pl. 154. 1775, non *Melastoma purpurascens* Swartz, 1788. *Melastoma purpurea* Willdenow, Sp. pl. 2(1): 590. 1799, nom. superfl. *Miconia purpurascens* (Aubl.) DC., Prodr. 3: 179. 1828.—TYPE: FRENCH GUIANA. "ad ripam rivuli in territorio Caux" [Kaw, 04°29′N, 52°20′W], Jun, *Aublet s.n.* (holotype: BM!).

Rhexia sileniflora Bonpland, Monogr. Melast. 2: 129, pl. 48. 1821. *Spennera sileniflora* (Bonpland) DC., Prodr. 3: 116. 1828. *Aciotis sileniflora* (Bonpland) Triana, Trans. Linn. Soc. London 28: 52. 1871.—TYPE: BRAZIL. Without locality, *Bonpland s.n.* (holotype: P, photo: F! NY!, fragment: F!).

Aciotis discolor D. Don, Mem. Wern. Nat. Hist. Soc. 4: 301. 1823.—TYPE: SAINT VINCENT: Cultivated plants (introduced from Guyana by Alexander Anderson and cultivated at the botanical garden), *Anderson s.n.* (holotype: G!, photos: F! GH! MICH! NY! S! US!, fragment: BR!; isotype: K!, photo: GH!).

Spennera fragilis DC., Prodr. 3: 115. 1828. *Aciotis fragilis* (DC.) Cogniaux in Eichler, Fl. bras. 14(3): 475. 1885.—TYPE: FRENCH GUIANA. Cayenne: "in umbrosis sylvaticis aquosis ad rivos Guianae," *Richard s.n.* (holotype: P; isotype: G!, photos: F! GH! NY!, fragment: F!).

Spennera pellucida DC., Prodr. 3: 115. 1828. *Aciotis purpurascens* var. *pellucida* (DC.) Cogniaux in Eichler, Fl. bras. 14(3): 474. 1885. *Aciotis pellucida* (DC.) Pittier et al., Cat. Fl. Ven. 2: 244. 1947.—TYPE: FRENCH GUIANA. [Cayenne?], *Richard s.n.* (holotype: P). [Synonymy according to Wurdack et al. (1993).]

Spennera grandifolia Miquel, Linnaea 18: 273. 1844.—TYPE: SURINAME. Without locality, *van Beeck-Vollenhoven s.n.* (holotype: U, digital image: MO!).

Spennera pellucida var. *subcordata* Crueger, Linnaea 20: 100. 1847.—TYPE: TRINIDAD. Without locality, *Crueger s.n.* (holotype: TRIN?).

Spennera uliginosa Bonpland ex Naudin, Ann. Sci. Nat. Bot., sér. 3, 14: 144. 1850. *Aciotis uliginosa* (Bonpland ex Naudin) Triana, Trans. Linn. Soc. London 28: 51. 1871.—TYPE: VENEZUELA. Amazonas: "in uliginosis secus ripas Orinoci," 1800, *Bonpland s.n.* (holotype: P, photo: F!, fragment: F!).

Spennera martinicensis Naudin, Ann. Sci. Nat. Bot., sér. 3, 14: 147. 1850. *Aciotis martinicensis* (Naudin) Urban, Symb. antill. 5: 446. 1908. *Aciotis discolor* var. *martinicensis* (Naudin) Stehlé, Bull. Soc. Bot. France 92(9): 265. 1945. *Aciotis purpurascens* var. *martinicensis* (Naudin) Stehlé, M. Stehlé & Quentin, Fl. Guadeloupe Depend. 2(3): 101. 1949.—TYPE: MARTINIQUE. "in insula Martinica," *Plée s.n.* (holotype: P, photo: F! NY!).

Aciotis circaeifolia var. *major* Cogniaux in Eichler, Fl. bras. 14(3): 473. 1885.—TYPE: BRAZIL. Pará: "ad Caripi juxta Para," Aug 1849, *Spruce 337* (lectotype, here designated: M!; isolectotype: NY!).

Aciotis fragilis var. *lancifolia* Steudel ex Cogniaux in Eichler, Fl. bras. 14(3): 475. 1885.—TYPE: SURINAME. "ad vias in distr. Para Surinameiae," Feb 1844, *Kappler 1680* (holotype: BR; isotypes: G! S!).

Aciotis purpurascens var. *longifolia* Cogniaux in Eichler, Fl. bras. 14(3): 474. 1885.—TYPE: FRENCH GUIANA. Without locality, *Leprieur s.n.* (holotype: BR; isotypes: G! W).

Aciotis longifolia var. *glabra* Huber, Bol. Mus. Paraense Hist. Nat. 4: 595. 1906.—TYPE: BRAZIL. Pará: "quebrada grande do Cerro de Canchahuaya," 13 Nov 1898, *Huber 1475* (holotype: MG, photo and fragment: F!).

Aciotis trinitensis Cogniaux in Urban, Symb. antill. 7: 310. 1912.—TYPE: TRINIDAD: "ad Carenage" [10°41′N, 61°36′W], 8 Jan 1908, *Broadway 2651* (holotype: BR, fragment!).

Herbs to subshrubs 0.5–1.5 (–2) m tall, erect; stolons present or absent; young stems quadrangular with conspicuous wings, glabrous or strigose with trichomes <0.5 mm long. Leaf blades 8–15 (–18.5) cm long, (2.5–) 3.4–6 (–6.5) cm wide, narrowly elliptic to ovate, apex acuminate, base attenuate to cuneate (rarely cordate), margin serrulate, membranous, 5-veined; adaxially green, glabrous or strigose with eglandular trichomes, abaxially green to purplish, glabrous or strigose with eglandular trichomes, without web-like striations; petiole 0.7–4 cm long, with evident wings, glabrous or strigose with eglandular trichomes <0.5 mm long. Inflorescence a thyrse, 10–18 cm long, 3.5–6 (–10) cm wide, with (3–) 5–6 pairs of first order paraclades; rachis glabrous; internodes 2–3.7 cm long; first order hypopodia 1–1.5 cm long, biparous; second order hypopodia 0.5–1 cm long, biparous or uniparous; third order hypopodia 0.4–0.7 cm long, biparous or uniparous; fourth order hypopodia absent to ca. 0.6 cm long, uniparous (rarely biparous); terminal paraclade (1) 2–4 cm long, with 3–4 (–5) flowers per cm; bracts 1–1.1 (–2) mm long, 0.2–0.4 mm wide, lanceolate, apex acute, abaxially glabrous. Hypanthium 2–3 mm long, 1–3 mm in diameter, terete, without filiform sclereids in parenchyma, abaxially and adaxially glabrous. Petals 3.1–4.5 (–5.2) mm long, 1–1.3 (–2.5) mm wide, elliptic, apex acuminate, base truncate, with a terminal glandular trichome, abaxially glabrous. Stamens with filaments 1.5–2.5 mm long, anthers 0.7–1 (–1.5) mm long, white, pedoconnective (0.2–) 0.3–0.6 mm long. Ovary without filiform sclereids in parenchyma, apex glabrous. Fruit baccate, pericarp and placenta caducous when ripe, 4–5 mm long, (2–) 3.5–5 mm in diameter, bilocular, pericarp membranous, green when ripe, translucent; seeds ca. 0.50 mm long, 0.30–0.40 mm wide. Figs. 3a, b, 4a, 5a, b, 6a, 26.

Phenology. Flowering and fruiting throughout the year.

Distribution (Fig. 27). Martinique, Trinidad, Costa Rica, Panama, and South America (except Chile, Paraguay, and Uruguay); flooded, dry, or periodically inundated areas, in primary forests and secondary vegetation; sea level to 1700 m.

Vernacular names. *Pegajosa* (Panama: Panamá); (Colombia: Amazonas; Witoto); *fruncida* (Colombia: Boyacá); *morilla* or *mora* (Colombia: Nariño); *jázirobe payachig colorado* (Ecuador: Napo); *cincak* and *daicáp cincák* (Peru: Amazonas); *arco-sacha, chirapa sacha, guirima-ey, muchaca, muchaquilla*, and *pia ong ky way* (Peru: Loreto); *ju'ika'a* (Brazil: Maranhão); *boyuyo morado* (Venezuela: Amazonas; *küdadei* and *tu'ñen utu' panëto' mechu' chan* (Venezuela: Bolívar); *bemindia wi, bémindja wi, bémindjawi sombowi, lebi suwawiwi, pata futu, sohouahouri, sombo wi*, and *sua wi* (French Guiana); *bèminja-wiwirie, witte swawiwiri*, and *zwawiwirie* (Suriname); *tiasakoreng*, and *wild sauari* (Guyana).

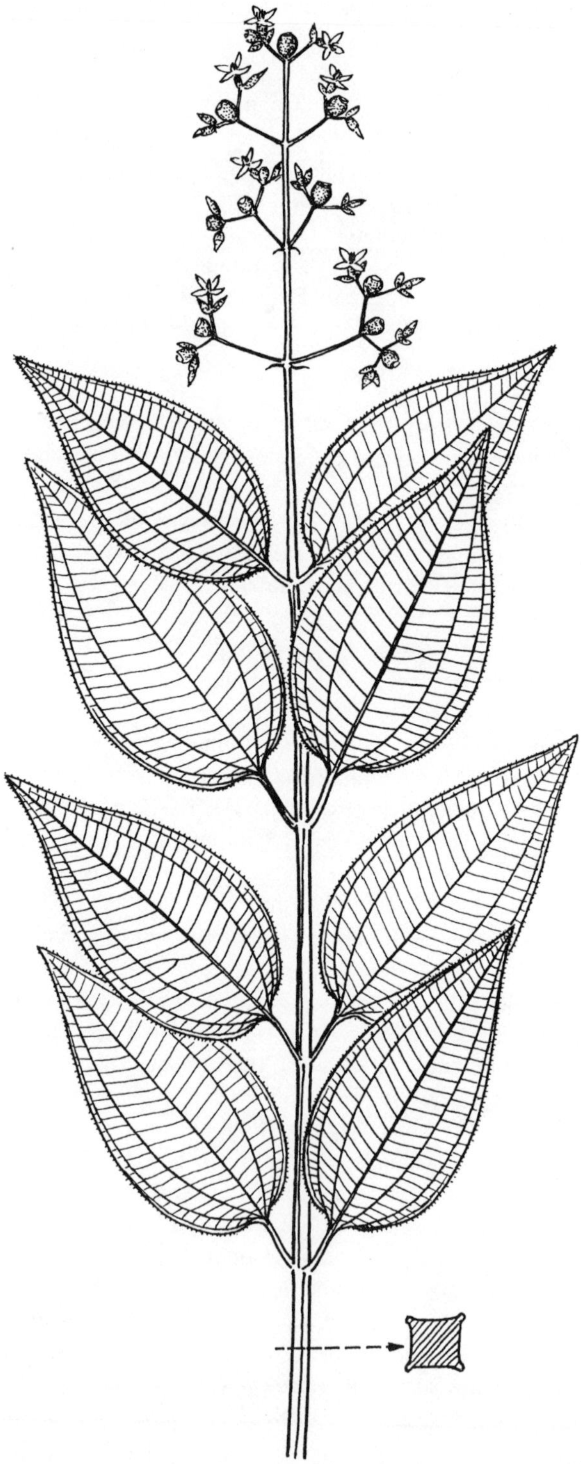

FIG. 26. *Aciotis purpurascens*. (Drawn by Bruno Manara; reproduced from volume 6 of *Flora of the Venezuelan Guyana* with the permission of the Missouri Botanical Garden.)

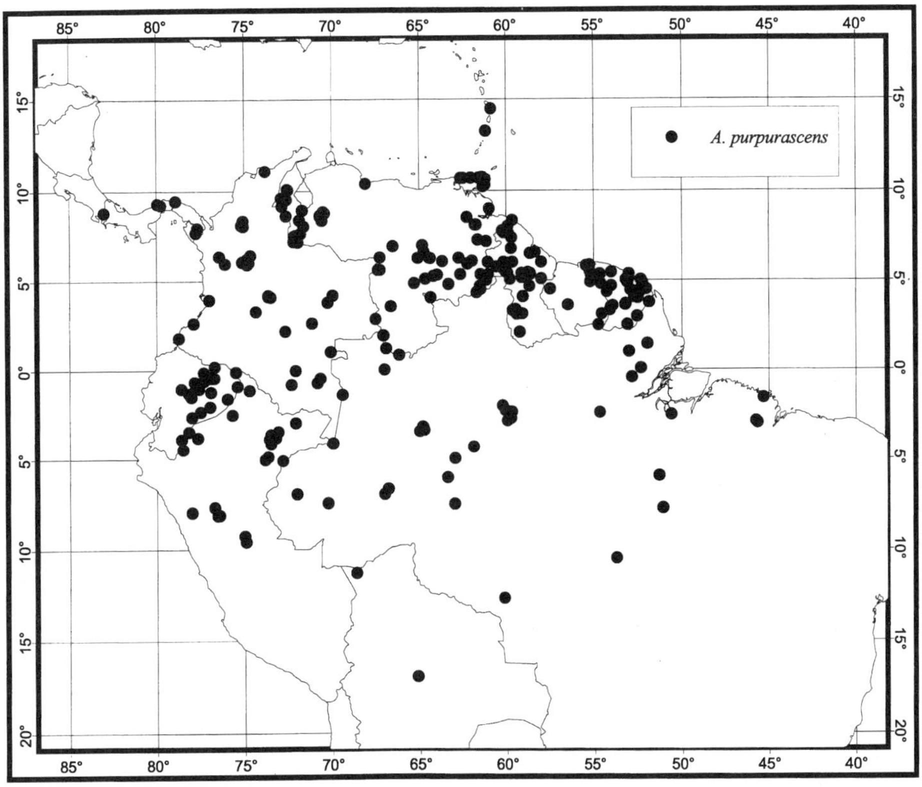

FIG. 27. Distribution of *Aciotis purpurascens*.

ADDITIONAL SPECIMENS EXAMINED. **Trinidad**. Morne Bleu, 10–12 mile, *R. O. W. et al. 12768* (K, TRIN); Morne Catherine, *Adams 14615* (TRIN); Waller Field, 10°37′N, 61°13′W, *Adams & Kalloo 13966* (TRIN); summit of Tucuche, 10°44′N, 61°25′W, *Alston 5110* (BM); El Tucuche Reserve, track to summit, 10°44′N, 61°25′W, *Barnard 501* (MO); Arima valley, *Berg 900* (S); Aripo Savanna, ca. 10°10′N, 62°00′W, *Britton et al. 277* (GH, NY, US); Mora forest, E of Sangre Grande, *Britton & Hazen 370* (GH, NY, US); Dibe Valley, *Britton et al. 1757* (GH, NY, US); Valencia, vicinity of Valencia Forest, *Britton & Britton 2092* (GH, NY, US); Carenage, 10°41′N, 61°36′W, *Broadway 2651* (BR); Aripo road, near the 4 mile post, *Broadway 5884* (BM, MO); Long Stretch, 23–24 mile post, *Broadway 7293* (BM); Arena Government Reserve, *Broadway s.n.* (F); Valencia, 1 mi SE of Valencia along E main road, *Davidse 2484* (MO, US); Santa Estrella, B.W.J., *Druce s.n.* (NY); Hollis Reservoir Road, ca. 1 mi downslope from the gates of Hollis Reservoir, *Harrimann 17587* (NY); Melajo Forest Reserve, *Johnson & Homer 271* (TRIN); St. George, summit of Blanchisseuse Rd., 1.75 mi W from Morne Bleu, *Proctor 21540* (IJ); Sangre Grande, ca. 3 mi SE, *Proctor 22442* (IJ); Mile 24 1/2 Valencia, Long Stretch, *Purseglove P.6451* (IJ, K, NY); Santa Isabella trail, near Simla, *Smith 10079* (NY, UC, US); Maraval, *Williams 10339* (TRIN); Four Rds, Tamana, *Williams 12870* (TRIN).

Costa Rica. PUNTARENAS: Las Cruces Tropical Botanical Garden, *Almeda et al. 3064* (CAS); Llanos de Coto, *Pittier & Tonduz s.n.* (BR, US). **Panama**. COLÓN: Canal Zone: Ft. Sherman, 2 mi SW of Ft. San Lorenzo, ca. 09°13′N, 79°45′W, *Antonio 1954* (CAS, MO); Chagres, Isthmus of Panama [09°19′N, 80°W], Mar 1850, *Fendler 149* (BM, F, K, US, W). —DARIÉN: alturas de Nique, Serranía de Pirre, above Cana mining camp, 07°40′N, 77°45′W, *Almeda & McPherson 5962* (CAS); Canaz–Río Setegantí, *Aranda et al. 1536* (CAS); S of El Real, alturas de Nique, near Cana mine, *McPherson 11503* (CAS, MO); Mamey, *Whitefoord & Eddy 349* (BM, MO).—SAN BLAS: Kuna, frente a Isla Niria Ubigandup, camino Digole, 09°26′N, 78°54′W, *Herrera 269* (MO).

Colombia. AMAZONAS: Araraquara, Trocha al Río Yarí, NE del Corregimiento, *Galeano et al. 883* (COL); Río Iagara–Paraná (affl. Río Putumayo), *Gasche & Desplats 82* (G, US); Río Caquetá, near Marímanteca, quebradón de Nejetei, *R. Jaramillo et al. 7990* (COL); Trocha entre Ríos Mirití–Paraná y Río Apaporis,

Centro Providencia, 03°14′S, 64°48′W, *P. Palacios et al. 405* (COL); Leticia, Km 11, ca. 04°10′N, 69°58′W, *Pinilla & Gutiérrez 14* (COL); Río Miritiparaná, Caño Guayacá, 00°30′S, 70°40′W, *Schultes & Cabrera 15775* (GH, US); Río Apoporis, Raudal Yayacopi (La Playa), *Schultes & Cabrera 16208* (GH, NY, US); Río Apaporis, between Kananarí and Pacoa rivers, *García-Barriga 13854* (NY, US).—ANTIOQUIA: Anorí, Corregimiento Providencia, La Tirana, 03°14′S, 64°48′W, *Albert de Escobar 301* (COL); Caucasia, along road to Nechí, ca. 19 km from Caucasia-Planeta Rica Rd., 08°02′N, 75°04′W, *Brant & O. Escobar 1243* (CAS, COL); Cáceres, Corregimiento El Doce, Bajo Cauca, *Callejas 449* (COL, HUA); Taraza, Corregimiento El Doce, 200 km NE de Medellín, camino de El Doce a Barroblanco, *Callejas et al. 1998* (HUA, NY, TEX); Bolívar, Vereda San Miguel, 10 km NE de Bolívar, Hda. Los Angeles, 05°56′N, 76°06′W, *Callejas & Fonnegra 7235* (COL, HUA); San Luis, Cañón del Río Claro, Parque Ecológico, *Cogollo & Borja 540* (COL, HUA, MO); San Francisco, Corregimiento Aquitania, *Fonnegra et al. 4294* (HUA, MO); Cocorna, Vereda La Piñuela, carretera a San Francisco, 06°02′N, 75°08′W, *Gómez et al. 280* (HUA, MO); Medellín–Bogotá, Sector Río Samaná–Río Claro, San Luis, *Hernández et al. 537* (COL, HUA); Medellín–Pto. Triunfo, near San Luis, 06°00′N, 75°00′W, *Juncosa 993* (US); Amalfi, *Lehmann 29* (BR, G, US); San José del Guaviare, La Libertad, 02°10′N, 72°40′W, *Marulanda 1205* (HUA); Ayapel, 08°19′N, 75°05′W, *Marulanda 2010* (HUA); Botero, Cauca Valley, *Pennell 10978* (NY); San Luis, Quebrada La Cristalina, 06°00′N, 74°45′W, *Ramírez & Cárdenas 1928* (HUA, MO); Mpio. Anorí, camino hacia Quebrada La Soledad, *Uribe 24* (HUA); San Carlos, Km 4.3 of road Juanes–Correg. Alto Samaná, 06°11′N, 74°52′W, *Zarucchi et al. 6736* (RB).—BOYACÁ: Casanare, Tauramena, orilla de quebrada Tauramenera, *Uribe 3592* (COL, US).—CALDAS: near to Victoria, *Uribe & Pérez-Arbeláez 2715* (COL).—CAQUETÁ: Solano, Araraquara, 00°50′S, 72°19′W, *M. V. Arbeláez & Sueroke 749* (HUA); Belen de los Andaquies–San José del Fragua, Km 8, *Bernal & Téllez 406* (COL); Florencia–Neiva, Km 34, *Cowan 2414* (US).—CHOCÓ: Camp Curiche–Q. Changame, 3.7 mi, *Duke & Idrobo 11519* (3) (US); Ensenada de Utria, 06°21′N, 76°26′W, *Echeverry 2213* (COL); Mpio. Pizarro, Km 30–33 de la carretera Pié de Pepé–Puerto Meluk, *Espina 1809* (CAS, COL, MO); Bahía Solano, Corregimiento El Valle, camino del Valle al Almejal, 06°21′N, 76°26′W, *Espina et al. 2464* (HUA, US); San José del Palmar, Hoya del Río Torito (afluente del Río Hábita), *Forero et al. 6948* (MO); Nuqui, near sea level, *Haught 5574* (COL, US).—GUAVIARE: San José del Guaviare, sector de laguna Pavón II, Territorio NUKAK, 02°36′N, 71°08′W, *Cárdenas et al. 7097* (MO).—MAGDALENA: Alto Río Buritaca, Alto de Mira, Cascada del Caño Negro 11°05′N, 73°48′W, *Madriñán & Barbosa 233* (MO).—META: Mesetas, Inspección de Policía La Uribe, puente sobre el Río Duda, carretera La Trece, 03°16′N, 74°21′W, *Betancur 1422* (HUA, NY); Santa Isabel, entre Restrepo y Cumaral, *García-Barriga 5014* (AAU); San Juan de Arama, Río Güejar, aterrizaje Los Micos, *Idrobo & Schultes 702* (COL, GH, MO, US); Villavicencio, 04°04′N, 73°30′W, *Karsten s.n.* (W); Raudal de la Macarena, Angostura No. 1, Río Guayabero, *Pinto et al. 339* (COL); Villavicencio, Cuenca del Río Ocoa, *Quiñones 2742* (COL); Cumaral, Vereda El Presentado, Caño La Tascosa, *Román 72* (COL); San Martín, Finca El Brasil, *W. R. Schwabe C 67/335* (COL).—NARIÑO: Río Sanquianga, Vereda Naidizales, Bosques de Guandal, *Caballero 166* (HUA); Guapi, Parque Nacional Natural Isla de Gorgona, camino entre Playa Blanca and Gorgonilla, ca. 02°36′N, 77°54′W, *Lozano & Rangel 5316* (COL, MO); Tumaco, *Romero-Castañeda 2709* (COL); Cabo Manglares, *Romero-Castañeda 3141* (COL); Tumaco, Chirambira, *Seemann 1034* (BM, K).—NORTE DE SANTANDER: Región del Sararé, Hoya del R. Cubugón entre El Caraño y El Indio, ca. 07°10′N, 72°10′W, *Cuatrecasas 13011* (COL, F, US); Bellavista, 09°09′N, 72°50′W, *Foster & Foster 1678* (A, COL, US); La Motilonia, Frontera Colombo-Venezolana, Hoya del Río de Oro, *García-Barriga & Lozano 18279* (COL, NY, US).—PUTUMAYO: Río Putumayo, Puerto Porvenir, arriba de Puerto Ospina, hacia La Loma, *Cuatrecasas 10658* (F, COL, US); Pepino–Mocoa old Rd, *Juajibioy 233* (GH); ca. 15 km of Puerto Asís, *King & Guevara 6116* (F, NY, US).—SANTANDER: Magdalena Valley, Campo Capote, 30 km E of Carare, *Gentry & Rentería 79999* (COL, MO); vicinity of Barranca Bermeja, between Sogamoso and Carare Rivers, El Centro, *Haught 2090* (COL, F, NY, US); Pto. Wilches–Pto. Santos, Km 16, *Killip & Smith 14834* (GH, NY, US); Aeropuerto, camino a Puerto Parra, Carare, Opón, Santander, *Rentería et al. 1529* (MO).—TOLIMA: Mariquita, *Uribe 2850a* (COL).—VALLE DEL CAUCA: Costa Pacífica, Quebrada de Guerré, afluente de quebrada del Chorro, Río Cajambre, *Arroyo 19* (MO); Río Digua Valley, between La Elsa and Río Blanco, *Killip 34741* (COL, NY, US); Bajo Calima, Vereda Campo Alegre, Río Bravo, ca. 03°55′N, 77°W, *Silverstone 557* (MO).—VAUPÉS: Alto Vaupés, Miraflores, *Gutiérrez et al. 756* (GH, NY); Mitú and vicinity, 01°00′N, 70°05′W, *Schultes & Cabrera 17838* (GH, NY, US); Río Vaupés, Circasia, 00°45′S, 70°50′W, *Schultes & Cabrera 19630* (NY, US). **Venezuela.** AMAZONAS: Depto. Casiquiare, alrededores de Yavita (Río Temi) y cerca de carretera Yavita–Pimichín, 02°50′N, 67°30′W, *Bunting et al. 3793* (US); Depto. Río Negro, Cerro Delgado Chalbaud, *Delascio 15370* (VEN); Atabapo, E del cerro Yapacana, 03°33′N, 66°38′W, *Guanchez 1116* (US); N end of large island just upstream from Neblina Base Camp along Río Mawarinuma, 00°50′N, 66°10′W, *Kral & Liesner 71837* (VEN); San Carlos de Río Negro, 3 km E forest, 01°56′N, 67°04′W, *Liesner 3490* (MO); Río Cuao, Río Orinoco, Territorio

Amazonas, *Maguire & Politi 27367* (F, NY, US); great rapids of the Orinoco, between Sanariapo Road and Salto Carestia (5 km N of Sanariego), *Maguire et al. 36149* (NY, RB); Río Guainia, *Maguire et al. 36430* (GH, NY); 1 km arriba de boca de R. Guayapo, tributario de R. Sipapo, *Morillo & Ishikawa 3480* (US); alrededores de San Simón de Cocuy, 1–2 km NO de La Piedra del Cocuy, 01°12′N, 66°54′W, *Morillo et al. 4075* (US); El Tobogán de la Selva, ca. 20 km S de Puerto Ayacucho, Dept. Atures, *Ruiz-Terán & Rodríguez 6015* (CAS, F, US); Depto. Atures, along road between dam site of Río Cataniapo and Gavilán, 35–40 km SE of Puerto Ayacucho 05°35′N, 67°15′W, *Steyermark et al. 122296* (MO, NY); ESE of Puerto Ayacucho, 10–30 km on road to Gavilán, 05°35′N, 67°22′W, *Steyermark et al. 113898* (US).—APURÉ: Dtto. Páez, selva de Cutufí, between Cutufí on the R. Cutufí and the R. Sanare, 07°09′N, 71°56′W, *Davidse & González 21928* (CAS, MO); Reserva Forestal San Camilo, Río Nulita, N del caserío San Camilo (El Nula), *Steyermark et al. 101319* (K, US).—BARINAS: San Isidro, 27 km de Barinitas, Dtto. Bolívar, 08°45′N, 70°25′W, *Aymard 2188* (MO, US); Dtto. Pedraza, trail El Algarrobo–Mesa de Canagua, E boundary of Parque Nacional Sierra Nevada, ca. 08°32′N, 70°39′W, *Dorr et al. 4718* (CAS, NY, US, VEN); Dtto. Bolívar, La Soledad, 08°49′N, 70°30′W, *Dorr et al. 7716* (CAS, NY); Ciudad Bolivia, fila del Alto de la Aguada, Mpio. Ciudad Bolivia, Dtto. Pedraza, 08°21′N, 70°34′W, *Ruiz-Terán 3438* (US); between La Esmeralda and El Curito, *Steyermark & Rabe 96523* (US, VEN).—BOLÍVAR: carretera El Dorado, Km 88, *Aristeguieta 3735* (VEN); Sifontes, near El Pilón and Sta. Elena de Uairén, ca. 04°40′N, 61°33′W, *Aymard 4577* (MO); Mpio. Foraneo Aripao, Araguaña, Alto Caura, ca. 04°74′N, 64°12′W, *Aymard & Delgado 6720* (US); Mpio. Raúl Leoni, a 85 km SSE de Entre Ríos, ca. 05°18′N, 63°59′W, *Aymard & Fernández 7258* (CAS, MO); Estac. Magdalena, El Palmar, 15 km NE, región del Río Grande, 08°05′N, 61°45′W, *Bernardi 7108* (G); región de los Ríos Icabaru, Hacha y cordillera sin nombre, a 280° de las cabeceras del Río Hacha, ca. 04°50′N, 65°18′W, *Bernardi 2719* (NY); Dtto. Cedeño, vicinity of Panare village of Corozal, 6 km from Maniapure toward Caicara, 06°55′N, 66°30′W, *Boom & Grillo 6356* (NY); Icarabú, 04°19′N, 61°44′W, *Croat 54134* (US); Pica Caicara del Orinoco–San Juan de Manapiare, Río Suapure; 202 km S de Caicara, ca. 07°S, 67°W, *Delascio & López 2767* (US, VEN); Tumeremo, San Martín de Turumban, Río Cuyuní, 07°15′N, 61°40′W, *Delascio & López 8747* (CAS); Río Venamo, Dtto. Rocío, SW de Anacoco, *Delascio & Reyes Lopez 8847* (VEN); Dtto. Cedeño, Morichal El Caballo, ca. 225 km at S of Caicara del Orinoco, 06°17′N, 67°15′W, *Holst & van der Werff 2560* (CAS, MO, NY, US); Piar, Río Aparamá, at rapids of Yuray-merú, tributary of Río Acanan, SW base of Amaruay-tepui, 05°55′N, 62°15′W, *Holst & Liesner 2658* (US); Dtto. Heres, alrededores del Helipuerto de C.V.G.-EDELCA Ichúm, 04°47′N, 63°21′W, *Huber 12360* (AAU, INPA, NY, US, VEN); alrededores de Sta. Elena, *Lasser 1645* (VEN); Caño Pablo, tributary of Río Caura, ca. 6 to 9 E km of Río Caura ca. 10 km ESE of Las Pavas (Salto Para), 06°14′N, 64°23′W, *Liesner & Morillo 13799* (NY); 17 km E of El Pauji, 64 km W of Santa Elena, Río Las Ahallas, 04°30′N, 61°30′W, *Liesner 19238* (US); Piar, 2 km W of west side of Amaruay-tepui, 2 km E of Río Acanán, 05°56′N, 62°16′W, *Liesner & Holst 20107* (US); Gran Sabana, ca. 10 km SW of Karaurin Tepu, 05°19′N, 61°30′W, *Liesner 23508* (MO, NY, US); Gran Sabana, San Ignacio de Yuruani, 05°00′N, 61°10′W, *Liesner 24296* (MO, NY); alrededores de Tumeremo, camino Tumeremo–Bolinche, entre Puesto (GN) y Caño Matuco del Río Negro, 07°15′N, 61°40′W, *Stergios et al. 3647* (CAS); Reserva Forestal Imataca, ca. 07°10′N, 61°10′W, *Stergios et al. 6024* (AAU); Las Trincheras, Reserva Forestal Río Caura, 06°58′N, 64°50′W, *Stergios et al. 8914* (MO, NY); along Río Caroní, near Urimán, 05°20′N, 62°40′W, *Steyermark & Wurdack 66* (F, NY); Quebrada O-paru-má, between Sta. Teresita de Kavanayén and Río Pacarairao (tributary of Río Mouak), *Steyermark 60444* (F, NY); E of Cerro El Picacho, N. of Las Nieves and Las Chicharras, 45 km N of Tumeremo, *Steyermark 89072* (NY); Cabeceras de Río Chicanán, 80 km (en línea recta) al SO de El Dorado, 06°05′N, 62°W, *Steyermark 89390* (NY); Río Nichare (afluente de R. Caura), near to R. Nichare, Sierra Maigualida y Sierra Cervatana, 06°15′N, 65°05′W, *Steyermark & Gibson 95663* (US); vecindades de Santa María de Erebato, Río Erebato, 05°05′N, 64°40′W, *Steyermark et al. 109874* (K, US); alrededores de Campamento Las Pavas, Salto Pava, 06°16′N, 64°27′W, *Tillett 763-99* (S); near El Palmar, ca. 08°05′N, 61°50′W, *Wessels Boer 2094* (US); Temblador, Medio Caura, *Williams 11673* (F, G, S, US); Río Villacoa (R. Auyacoa), Salto de Humito, *Wurdack & Monachino 41158* (GH, NY, R, RB, US).—CARABOBO: ca. 18 km to 20 km S of Puerto Cabello, S of San Esteban, 10°23′N, 68°04′W, *Liesner & Medina 13742* (MO, US, VEN).—DELTA AMACURO: E de Río Grande, ENE de El Palmar, *Blanco 162* (MO, NY, US); Tucupita, 5–14 km ESE of Los Castillos de Guayana, 08°31′N, 62°17′W, *Davidse & González 16401* (MO); Río Amacuro, Venezuela-British Guiana frontier, Sierra Imataca, *Steyermark 87184* (NY); Depto. Antonio Díaz, along Caño Joba–Saburu, W of Caño Guayo and E of Caño Sacupana, 08°59′N, 61°00′W, *Steyermark et al. 115160* (MO, VEN).—MÉRIDA: entre el Quinó y el R. Socopó, Mpio. El Morro, Dtto. Campo Elías, *López-Palacios 2722* (US); road to Mijagual–Los Mesones, from 1 km S of Puente Victoria, Cordillera de Mérida, Dtto. Tovar, *Tillett 738-563* (US, VEN).—MONAGAS: cerca de Carifito, *Aristeguieta 3983* (MO, NY).—SUCRE: Cerro Sacamanteca, NO de Yoco, *Berti et al. 1283* (BR, MO, US); Península de Paria, summit of Cerro Patao, 10°41′N, 62°03′W, *Milliken et al. 56* (MO); entre Manacal y Los Pocitos de Sta.

Isabel, 25 km NO de Irapa, 10°40′N, 62°38′W, *Morillo 2544* (F); Península de Paria, Cerro Espejo, entre Manacal y Pauji, ca. 10°50′N, 61°50′W, *Steyermark & Rabe 96048* (NY).—TÁCHIRA: 5 km SW of Dam site on Río Caparo, 31 km ESE of St. Barbara, *Liesner & González 9305* (MO); around Represa Dorada, *Liesner & González 10372* (MO, NY, US); La Buenana, 6–12 km W of Quebrada Colorado, ca. 35 km SSE of San Cristóbal, 07°28′N, 72°09′W, *Liesner & González 10784* (MO); Dtto. Uribante, Sector El Cedral, 08°01′N, 71°37′W, *Rivero et al. 1883* (US); Cerro of Cuchilla La Pabellana, W of San Joaquina de Navay, 07°37′N, 71°47′W, *Steyermark et al. 119426* (MO, US); Ayarí, 07°32′N, 71°53′W, *Steyermark et al. 119496* (MO); Dtto. Uribante, ca. 3 km outside Sibera along old road to Pregonero, savannas near Siberia, 08°55′N, 71°41′W, *van der Werff & Ortega 8110* (MO, NY).—ZULIA: carretera Machiques–Colón, 10°04′N, 72°32′W, *Aristeguieta et al. 6782* (NY, US); Dtto. Perijá, alrededores de la Estación Hidrológica Aircuaisa–Pie de Monte, 09°35′N, 72°53′W, *Bunting et al. 10865* (NY, US); Dtto. Catatumbo, Machiques–La Fría road, ca. 08°36′N, 72°36′W, *Bunting & Alfonzo 13274* (NY, US); between Río Tokuko and Río Aricuiza, carretera Machiques–Colón, about 90 km S, *Steyermark & Fernández 99607* (COL, K, US); Sierra de Perija, S of Misión de los Angeles de Tokuku, ca. 10°00′N, 72°30′W, *Steyermark 99967* (K, US). **Guyana**. Pomeroon, Moruka River, Huradaiah, *de la Cruz 1021* (GH, NY, US); Bartica, on the Essequibo river, 58°37′N, 06°25′N, 58°37′W, *de la Cruz 1920* (F, GH, MO, NY, US); vicinity of Wismar, on the Demerara River, 06°00′N, 58°00′W, *de la Cruz 2468* (F, GH, MO, NY, US); Pomeroon river, Pomeroon district, *de la Cruz 2936* (F, GH, K, MO, NY, US); northwest, Waini River, 08°20′N, 59°40′W, *de la Cruz 3709* (F, GH, MO, NY, US); Omai, between Essequibo River and first clearing, 05°26′N, 58°46′W, *Gillespie & Persaud 1600* (NY); Aurora, 0–1 km S of camp, 06°47′N, 59°44′W, *Gillespie & Tiwari 2119* (C, INPA, MO, NY); Mazaruni River, between Imbaimadai and Iruwa River, 05°42′N, 60°17′W, *Gillespie et al. 2941* (MO, NY); Tumatumar, *Gleason 926* (GH); Essequibo Islands-W. Demerara, McNabb Conservancy Area, W coast Essequibo, *Harris TP 390* (K); Yarikita Police Station, junction of Yarikita and Amakuru Rivers, 08′N, 59°55′W, *Hitchcock 17653* (GH, US); Essequibo Islands-W. Demerara, Arawak Amerindian land, 3 km up Pokerero River, 06°34′N, 58°21′W, *Hoffman & Capellaro 901* (INPA, TEX); Pakaraima Mts, upper Karowrieng River at Maipuri Falls, 05°41′N, 60°13′W, *Hoffman 2938* (US); Pakaraima Mts, NW side of Mt. Ayanganna, 05°24′N, 59°57′W, *Hoffman & Henkel* 3273 (INPA, TEX); Upper Demerara-Berbice, Berbice River, 230 km upriver from mouth, N of Kwakwani, 05°04′N, 58°00′W, *Hoffman & Aymard 4004* (US); Kanuku Mts, 03°08′N, 59°23′W, *Jansen-Jacobs et al. 396* (AAU, COL, K, MO, NY, US, WIS); Rupununi district, Kuyuwini River, 02°5′N, 59°15′W, *Jansen-Jacobs et al. 2854* (NY), *Jansen-Jacobs et al. 3096* (NY, US); U. Takutu-U. Essequibo, SE Kanuku Mts, Crabwood Creek, 03°07′N, 59°06′W, *Jansen-Jacobs et al. 3392* (MO); Mt. Russell District, *Jenman 2094* (K); Mattews Ridge, Nortwest District, *Maguire 40500* (NY, US); Macaw Creek to Holitipu camps, Kartabo Mtn, 06°00′N, 61°05′W, *McDowell & Gopaul 3083* (NY); entrance Mabura Hill-Demerara Woods, 05°22′N, 58°38′W, *McDowell 3404* (NY); Potaro–Siparuni, SW base of Eagle Mt, along Potaro–Konawaruk Rd, 05°12′N, 59°07′W, *McDowell & Gopaul 3519* (NY); Eping River, 06°00′N, 60°10′W, *McDowell & Stobey 3873* (NY); Bartica–Potaro rd, 24 mi S of Bartica, at Forest Service's 24 Mile Camp, *Mori et al. 8086* (NY); West Demerara Region, Mabura Hill, 05°20′N, 58°40′W, *Pipoly & Boyan 7591* (NY, US); Barima–Waini, head of Barima River, 15 mi E of Arakaka, 2 mi S of Barima R., ca. 2 mi of Tenapu Cr., 07°37′N, 59°53′W, *Pipoly & Lall 8128* (US); along the Koatse River, ± 2 km E of Pong R., 5 hrs walk S of Chinoweing Village, 05°28′N, 60°04′W, *Pipoly et al. 10794* (NY, US); Ayanganna Plateau, ± 2 km W of base camp in Koatse River Valley, 05°28′N, 60°04′W, *Pipoly et al. 10985* (NY, US); Mazaruni River, 1.25 mi below its junction with the Kamarang River, 05°45′N, 60°35′W, *Robinson 85-0060* (US); Essequibo River, Moraballi Creek, near Bartica, *Sandwith 545* (K, NY); Kanuku Mts, Moco Moco R., 03°18′N, 59°39′W, *Snow 32* (K); Mazaruni–Potaro, Potaro road, 05°23′N, 59°08′W, *Stockdale 8807* (K); West Demerara, Mabura Hill, 180 km SSE of Georgetown, near Mabura, 05°20′N, 54°40′W, *Steege ter et al. 308* (CAS, K); Kariako Village, Barama River, 07°22′N, 59°42′W, *van Andel et al. 726* (US); Potaro river, Kangaruna, *Whitton 58* (K). **Suriname**. Tumuc Humac Mts, Talouakem, 02°29′N, 54°45′W, *Acevedo-Rodríguez et al. 5937* (US); Paka paka (Saramacca R.), Ebbatop, *Florschütz & Florschütz 1519* (C, NY); Frederik Top, 3.5 km SSE of Juliana Top, 03°36′N, 56°30′W, *Irwin et al. 54919* (F, NY, S, WIS); Paramaribo, 10 km S, Helena Christinaweg, line B, 05°50′N, 55°15′W, *Jonker-Verhoef & Jonker 210* (AAU), Brokopondo, Brownsberg, 04°53′N, 55°13′W, *Koster 13032* (AAU, US); Jodensavanne–Mapane area, 05°05′N, 55°04′W, *Kramer & Hekking 2453* (IJ, US); Libanonweg, S of Paramaribo, ca. 05°20′N, 55°12′W, *Kramer & Hekking 2780* (NY); Tibiti savanne, *Lanjouw & Lindeman 1721* (K, NY); Marowijne, Nassau Gebergte, 04°48′N, 54°36′W, *Lanjouw & Lindeman 2883* (NY); Lely Mts, *Lindeman et al. 10* (COL, US, Z); Kabalebo Dam Project, distr. Nickerie, 04°30′N, 57°30′W, *Lindeman & Roon de 879* (US); Tafelberg, *Maguire 24827* (NY); Marowijne River, *Palmer-Jones 68* (K); Tapanahoni River, *Rombouts 623* (COL, K, MO, NY); Litani, Taponté, *Rombouts 759* (K, NY); Amaka kondre, *Sauvain 697* (CAY). **French Guiana**. Massif Lucifer, *Billiet & Jadin 1639* (BR, CAY); Bassin du Sinnamary, Piste de Saint-Elie, Km 16, 05°20′N, 53°00′W, *Bitaillon 103* (US); Cayenne–St.

Laurent, *Cowan 38911* (K, NY, US); Moengo, *Cowan 38962* (NY, US); Sure le Tampoc, Saut Koumakou Soula, *Cremers 4464* (CAY, US); SO Cayenne, *Cremers 5242* (CAY, US); Montagne Caco, S. de Cayenne, 04°30′N, 52°20′W, *Cremers 6927* (CAY, US); Riv. Mana, Saut Dalles, *Cremers 7237* (CAS, CAY); route de Bélizon, R.N.2, 00°04′N, 52°20′W, *Cremers 10038* (CAY, NY, US); station des Nouragues–Bassin de l'Approuague–Arataye, 04°03′N, 52°42′W, *Cremers et al. 10946* (CAY); Camp Eugène–Bassin du Sinnamary, 04°51′N, 53°04′W, *Cremers & de Granville 13715* (CAY); Grand Santi, 04°20′N, 54°15′W, *Daniel R 33* (CAY, US), *Daniel 45* (CAY); environs de Saint Laurent du Maroni, à 9 km sur la route de Saint Jean, *Descoings & Luu 20374* (CAY); Grégoire, *Deward 193* (CAY); Antouka–Bassin du Maroni–Lawa, 03°29′N, 54°01′W, *Fleury 233* (CAY); Saut Verdum, Rivière Petit Ouaqui, *de Granville 1911* (CAY, US); Orapu, Route de l'Est, *de Granville 2775* (CAY, US); Massif des Emerillons, *de Granville 3767* (CAS, CAY, US); Montagne de Kaw, foret sur cuirasse lateritique a proximite du camp Caiman, *de Granville 4178* (CAY, US); Haute Camopi, Mont Belvédère, 02°30′N, 53°05′W, *de Granville 7038* (CAY, US); Mont Atachi Bacca–Region de l'Inini, 03°33′N, 53°55′W, *de Granville 10868* (CAY); Gruigel–Saut Macaque, fleuve Ouaqui, 05°05′N, 53°W, *de Granville B.4887* (CAY); without locality, 1802, *Gabriel s.n.* (G); Zidok ville, *Grenad 200* (CAY); Saül, Monts La Fumée, 03°40′N, 53°10′W, *Hahn 3596* (US); piste de St. Elie, CD–21, Km 15.7, 05°00′N, 53°10′W, *Hahn 3756* (US); Montagne du Mahury–Ile de Cayenne, 04°52′N, 52°15′W, *Hoff 5416* (CAY, RB); Saut Stephanie–Bassin du Sinnamary, 04°27′N, 52°54′W, *Hoff 5990* (CAY); La Mana, coll. 1823–24, *Leschenault s.n.* (US); Maroni, *Mélinon s.n.* (US); vicinity of Eaux Claires, on the route to Bélizon 0–4 km S of Eaux Claires, 93°37′N, 53°12′W, *Mori et al. 22072* (CAY); Oyapock fleuve, sure la crique Gabaret, 04°N, 52°30′W, *Oldeman B-1559* (CAY, US); Ouanary, crique, pres de la montagne des trois Pitons, 04°03′N, 52°42′W, *Oldeman B-2637* (CAY, NY); Approuague, Crique Fromager, base crique Courouaie, *Oldeman B-2654* (CAY); Petit Saut, Km 16, 05°20′N, 53°W, *Prévost 2065* (CAY, K); St. Laurent du Maroni, *Sastre & Sastre 301* (CAY, US); St. Laurent–Paul Isnard, Km 10 SE, Sosakaba, 5–15 m, 05°25′N, 54°00′W, *Skov et al. 7428* (CAY, TEX, US). **Ecuador**. MORONA-SANTIAGO: Centro Shuar Yukutais, 03°30′S, 78°10′W, *Bennett & Gómez 3519* (QCNE, TEX); Chupiantza, Cordillera Cutucú, 02°40′S, 78°00′W, *Camp E-1035* (NY, US); valley of the Ríos Negro and Chupiantza, on the trail from Sevilla de Oro to Méndez, El Partidero, *Camp E-1527* (NY); Taisha, ca. 02°23′S, 77°30′W, *Cazalet & Pennington 7755* (K, NY, US); Méndez, 02°23′S, 77°30′W, *Harling 964* (NY, S); Yurupaza, *Harling 985* (NY).—NAPO: Tena, ca. 01°04′S, 77°36′W, *Asplund 9322* (CAS, G, R, S, US); Coca–Lago Agrio Rd, Km 5, Lorocachi, 00°24′S, 76°59′W, *Brandbyge et al. 30201* (AAU, K, QCA, RB); Misahuallí–Hotel Jaguar, 1 hour down Río Napo, *D'Arcy 14027* (MO); Coca–Archidona, Km 30, vía Hollín –Loreto, ca. 00°45′S, 77°40′W, *Freire-Fierro & León 2751* (MO, QCA); Puerto Napo–Puyo rd, Km 3, 01°03′S, 77°47′W, *Freire-Fierro & León 2755* (MO, QCA); Jatun Sacha Biological Reserve, 8 km from Pto. Misahuallí, 01°04′S, 77°36′W, *Freire-Fierro 2838a* (MO, QCA), *Freire-Fierro 2840* (MO, QCA); Río Bueno–Santa Rosa trail, *Harling et al. 7187* (GB, MO, US); Coca, lower Río Payamino, 00°56′S, 75°24′W, *Harling et al. 7761* (GB); Cañón de los Monos, ca. 12 km N of Coca (Puerto Francisco de Orellana), *Harling & Andersson 11745* (AAU, GB, MO, US); Coca, vía los Aucas, 00°28′S, 76°44′W, *J. Jaramillo & Rivera 166* (AAU, NY, QCA); vía Lago Agrio–Río San Miguel, frontera con Colombia, 00°10′N, 76°42′W, *J. Jaramillo 2231* (GB, QCA); Río Zumino, tributary of the R. Napo, 5 km NE from Sta. Rosa, *Lugo 187* (F, GB, MO, NY, US, WIS); Coca (Pto. Francisco de Orellana)–Lago Agrio Rd, Km 3, ca. 00°24′S, 76°59′W, *Lugo 2935* (F, GB, NY, S, US); 21 km E of Loreto, 00°38′S, 77°19′W, *Renner 69323* (AAU, CAS, QCA, TEX).—PASTAZA: Mera, ca. 01°20′S, 78°10′W, *Asplund 18317* (S); Lorocachi, 01°38′S, 75°58′W, *Brandbyge & Asanza 30804* (AAU, QCA, US); Mera–Río Anzu, km. 11, 01°20′S, 78°10′W, *Freire-Fierro 2786* (MO, QCA); Madretierra–La Isla, Coll. in La Isla, 01°32′S, 78°02′W, *Freire-Fierro 2794* (MO, QCA); Auca–Coca Rd, 115 km S, 10 km al S de la frontera Napo–Pastaza; cerca del Río Tigüino, Carretera de PETRO-CANADA en construcción, 01°15′S, 76°55′W, *Hurtado et al. 1323* (G, TEX, WIS); Cuansha–Nalpi, *Lugo 1447* (AAU, GB, MO, US); Canelos, *Lugo 1560* (GB, K, NY, S, US); Río Capahuari, ca. 12 km N of Sarayacu, *Lugo 4113* (AAU, GB, MO, US); Río Chullana, ca. 15 km N of Sarayacu, ca. 01°44′S, 7°29′W, *Lugo 4147* (A, F, GB, US, WIS); El Porvenir, ca. 5 km N of Puyopungo, *Lugo 4923* (A, F, GB, US, WIS); Sarayacu–Copataza, Km 10, *Lugo 5529* (GB, US); Shiguacocha, 4 km E of Sarayacu, *Lugo 5562* (GB, MO); Montalvo, along Río Bobonaza, 0–1 km N of the military camp, 02°05′S, 76°58′W, *Øllgaard & Molau 13247* (AAU, GB, QCA).—PICHINCHA: road Sto. Domingo–Rosa Zárate (Quinindé), Río Blanco, *Harling et al. 9324* (GB).—ZAMORA-CHINCHIPE: ca. 4 km E of Paquisha, 03°55′S, 78°35′W, *Øllgaard et al. 90409* (MO, QCA, QCNE). **Peru.** AMAZONAS: Huampami, 04°30′S, 78°30′W, *Berlin 533* (GH); Condorcanqui, Río Cenepa region, E of Huampami, *Berlin 1506* (WIS, XAL); Quebrada Basusinuk, tributary of Huampami, *Kayap 27* (GH); Bagua, Montenegro–Chiriaco, *Sagástegui 5908* (US).—HUÁNUCO: Pachitea, Puerto Inca to Sungaro, 09°18′S, 75°00′W, *Foster 8751* (US); SE of Pucallpa, next to junction of Río Pachitea and Yuyapichis, Panguana village, 09°37′S, 74°56′W, *Morawetz & Wallnöfer 112-241085* (US).—JUNÍN: Puerto Vessup, *Killip 26229* (NY, US).—LORETO: Iquitos, 03°51′S, 73°13′W,

Asplund 14072 (S); Maynas, Tamishiyacu, *Ayala 2877* (MO, NY); Río Yuvineto, affluent of Río Putumayo, Secoya Indians, *Barrier 249* (G); 7.5 km NE of Río Nanay, along road to Puerto Almendra, 7 km from Río Nanay, *Croat 18325* (XAL); Iquitos–Río Itaya, ca. Km 6, *Croat 19068* (MO, WIS); Maynas, Zungaro Cocha, near Iquitos, *Dodson & Torres 2917* (F, MO); Alto Amazonas, Pongo, cerca a Yurimaguas, *Ferreyra 17466* (US); Requena, Genaro Herrera, Río Ucayali, *Gentry et al. 21353* (MO, US); Mishana, Río Nanay, halfway between Iquitos and Sta. María de Nany, 03°45′S, 73°20′W, *Gentry et al. 36434* (AAU, CAS, MO, NY, TEX); Yurimaguas and Balsapuerto, lower Río Huallaga basin, 03°50′S, 73°15′W, *Killip & Smith 28173* (NY, US); Mishuyacu, near Iquitos, 03°50′S, 73°13′W, *Klug 144* (F, NY, US); Florida, Río Putumayo, 00°08′S, 75°30′W, *Klug 2174* (BM, F, G, GH, K, MO, NY, S, US, WIS); Yurimaguás, Huallaga, *Kuhlman 1363* (RB); Lagunas, trail to Argentina, *McDaniel & Rimachi 16444* (F, MO, US); Río Tigre, 02°33′S, 75°42′W, *McDaniel & Rimachi 18439* (MO, NY, US); Iquitos–Nauta Rd, Km 6 from Quistococha to Varillal, *Rimachi 7450* (BR); Gamitanacocha, Río Mazán, 03°30′S, 73°04′W, *Schunke 146* (F, GH, NY, US); C. Portillo, Inchua, 08°00′S, 78°00′W, *Soukup 3021* (F); Pebas, Bora Native community of Brillo Nuevo, Yaguasyacu River, 03°00′S, 72°05′W, *Treacy & Alcorn 20* (WIS); Maynas, Iquitos, Allpahuayo, Estación Experimental del Instituto de Investigaciones de la Amazonía Peruana (IIAP), *Vásquez & N. Jaramillo 14450* (MO, TEX); Lower Río Nanay, *Williams 334* (F); Pinto–Cocha, on Nanay River, *Williams 803* (F, NY); La Victoria on the Amazon River, Peru/Brazil, *Williams 2568* (F, NY); Alto Río Itaya, *Williams 3209* (F, NY); Lower Río Huallaga, *Williams 4739* (F, NY).—SAN MARTÍN: Río Shilcayo, valley of San Martín, 1–4 km NE of Tarapotó, *Belshaw 3294* (F, GH, LL, MICH, NY, US, WIS); Prov. Mariscal Cáceres, 12 km W of Tocache Nuevo, near Río Tocache, 08°10′S, 76°32′W, *Gentry et al. 25652* (MO, US); camino a Pushurumbo, 5–6 km de Palo Blanco, Mariscal Cáceres, Tocache Nuevo, *Schunke 5557* (F, K, NY, US, W); camino a Shunté, *Schunke 7345* (MO, XAL); antigua carretera a Limón, cerca a la Granja Santa Yves, *Schunke 10285* (US); Tarapoto, *Woytkowski 35084* (F, G, MO). **Bolivia.** COCHABAMBA: Chapare, Chimoré, 16°59′S, 65°08′W, *Badcock 360* (K).—PANDO: Nicolás Suárez Cobija, cerca de El Porvenir, 11°02′S, 68°44′W, *Beck 17080* (US). **Brazil.** AMAPÁ: Igarapé Carnot Pequeno, E da estrada nova Calçoene, Oiapoque, ca. 03°45′N, 51°52′W, *Austin et al. 7360* (MO, NY, US); Macapá–Mapatí Rd, *Black & Lobato 50-9731* (NY); tributary of Igarapé do Casino, at junction of trails to Rio Araguary and Serra do Viada, *Cowan 38371* (NY); Rio Amapari, Igarapé Pedra Preta, *Cowan 38510* (NY, RB, US); Rio Aguari, 01°26′N, 51°58′W, *Pires et al. 50805* (F, NY, US); Aldéia Galibi–Oiapoque, *Rabelo & Non 1498* (MG).—AMAZONAS: Lago de Castanho–Mirim, estrada de Petrobras, *Albuquerque et al. 863* (INPA); vila Bittencourt, Rio Japurã, Serrinha, 01°25′S, 69°25′W, *Amaral et al. 517* (INPA, NY); Manaus–Porto Velho highway, Km 215, *Campbell et al. P20781* (INPA, K, NY, US); fronteira de Brasil e Colombia, Mpio. Japurá, confluência dos rios Traíra e Apaporis, próximo a cachoeira Urumutum, 03°14′S, 64°48′W, *Cid Ferreira & Lima 3711* (F, INPA, MO, NY); Mpio. Mancio Lima, estr. para o lugar Barão entre os Km 30 e 52, 07°S, 72°W, *Cid Ferreira et al. 5238* (NY, US); Manaus–Boa Vista road (BR-174), Km 115, *Farney et al. 424* (INPA); Taperinha bei Santarém, *Ginzberger et al. 603* (W); Paricatubal, Rio Purus region, Lago Agua Fria, 04°22′S, 61°55′W, *Gottsberger & Döring 12-17286* (MO); Tapaua, Rio Ipixuna, 29 km upstream from Tapaua, 06°05′S, 63°25′W, *Gottsberger & Döring 12-18186* (MO); CEPLAC, Km 30 on road from Manaos–Itacoatiara, *Lasseigne P22586* (INPA, MG, NY, US); 50 Km NE of Manaus, Road ZF-1, 3 km off Itacoatiara road, 02°45′S, 59°45′W, *Lowe et al. 4120* (INPA, K); Manaus–Humaitá, Km 620, *Lowrie et al. 83* (INPA, NY, US); Taracuá on Rio Uaupés, 06°43′S, 66°47′W, *Melin 103* (S); Manaus–Itacoara Rd, Km 118, 02°19′S, 60°5′W, *Monteiro et al. 4* (INPA); Rio Cuieras, Igarapé Lobishomen, ca. 02°03′S, 60°15′W, *Mori & Gracie 21813* (INPA, NY); ca. 80 km N de Manaus, Distrito Agropecuario da Suframa, Fazenda Esteio, 02°26′S, 59°48′W, *Nee 42826* (MBM, NY); Tefé, Mouth of Rio Bauana, ca. 03°29′S, 65°00′W, *Plowman et al. 12558* (NY, US); N from São Gabriel, along BR 307, near Igarapé Freitas and at Equator; 00°00′S, 67°00′W, *Poole 2032* (INPA, NY); Basin do Rio Purus, Rio Cunhuá at Deni Indian village, 06°43′S, 66°47′W, *Prance et al. 16459* (F, GH, INPA, M, R, S, US); Manaus, Igarapé da Sto. Antonio, *Rodrigues & Chagas 1951* (INPA, MG, US); Reserva Forestal Ducke, Manaus–Itacoara, Km 26, Igarapé do Tinga, 02°53′S, 59°58′W, *Sothers 1004* (INPA); Serra dos Carajás, 1–4 km along road from camp Azul toward AMZA camp N-1, 06°06′N, 50°17′W, *Sperling et al. 5868* (NY); Rio Negro, Santa Isabel, *Tate 103* (NY); Humaitá–Labrea Rd, Km 63, 07°33′S, 63°01′W, *Teixeira et al. 903* (INPA, MG, NY, US); Manaus–Caracaraí, Km 44–45, da BR 174, *Vasconcellos & de Carvalho 10329* (UEC).—CEARÁ: without locality, *Allemão & de Cysneiros 626* (R).—MARAHHÃO: Carutapera, basin of the Rio Maracaçumé, Ka'apor Indian Reserve, *Balée 1033* (K, MA, MG, NY, US); Forest S of Km 157 of BR 316, Araguaná, 03°00′S, 45°40′W, *Daly et al. D230* (INPA, NY, US); Rio Alto Turiaçu, Nova Esperança, 02°55′S, 45°45′W, *Jangoux & Bahia 260* (MG, NY); Turiaçu, Km 6 da BR 106 Maracaçumé–Sta. Helena, fazenda Maracaçumé Agro Industrial Grupo Mesbla, 02°55′S, 45°45′W, *Rosa & Vilar 2831* (NY, US).—MATO GROSSO: Xavantina–Cachimbo Rd, Fazenda Bonsucesso, Margen da BR 80, 10°33′S, 53°44′W, *Passos 1008* (HRB, US); Aripuanã, Projeto Juína, *Silva & Rosario 4901* (INPA, MG).—PARÁ: Tumuc–Humac, via Rio

Cumina, Cachoeira do Tronco, *de Sampaio 4999* (R); Capitão Poço, colonia São José, *Elias 270* (HB); Reserva Forestal de Gorotire, Río Fresco, 07°47′S, 51°07′W, *Gottsberger & Posey 117–20183* (US); near mouth of Rio Embira, tributary of Rio Tarauaca, 07°30′S, 70°15′W, *Krukoff et al. 5074* (A, G, K, NY); Belém–Brasilia Highway, Km 105, *Maguire et al. 56054* (K, NY, US); Almeirim, Mt. Dourado, 00°27′S, 52°50′W, *Pires & Silva 1541* (INPA); Rios Pacaja and Muirapiranga, 02°33′S, 50°38′W, 0–5 km N of Rui Muirapiranga, *Prance et al. 1586* (F, G, NY, S, US, Z); Km 64, vicinity of Piritoro, Road BR-22, Capanema to Maranhão, *Prance & Pennington 1960* (NY, US); Rio Jarí, estrada do Munguba, Km 14, ca. 01°N, 53°W, *Silva 2197* (NY, US); Santarém, Km 35 da estrada do Palão, igarapé Curupira, 02°26′S, 54°41′W, *Silva & Souza 2221* (NY, US); Rio Vermelho, BR-163, Cuiabá-Santarém highway Km 867, Serra do Cachimbo, *Silva et al. 208* (INPA, NY, US); Caripi, juxta Para, *Spruce s.n.* (BM, G, W); Benjamín Constante, *Whiffin 11* (K).—RONDÔNIA: 3–4 km SW of Mutumparaná on railroad to Abunã, *Prance et al. 5442* (INPA, NY); a 29 km de Vilhena, 12°45′S, 60°10′W, *Vieira et al. 696* (GH, INPA, MICH, MG, NY, US).—RORAIMA: base of Serra Tepequém, Boca da Mata, *Prance et al. 4322* (INPA, NY, US); Posto Mucajaí, Rio Mucajaí, *Prance et al. 11172* (F, GH, INPA, M, NY, R, S, US); Auaris, 04°03′N, 64°22′W, *Prance et al. 9856* (F, INPA, S, US); Serra das Surucucús, *Ribeiro 614* (HRB).

Aciotis purpurascens is easily recognized by its winged stems, thyrsoid to paniculate inflorescences, and baccate fruits. Occasionally, it grows as an epiphyte (*Cremers 7592, 7237*; pers. obs.), perhaps a result of dispersal by birds. The leaves vary from glabrous to strigose throughout its extensive range. This plasticity has been recognized taxonomically, as reflected by the many names included in the synonymy; however, study of the material now available shows that the regional variants all fall within the circumscription of *A. purpurascens.*

Aciotis purpurascens shares with *A. rubricaulis* conspicuous wings, thyrsoid inflorescences, and baccate fruits; however, *A. rubricaulis* has mainly densely pubescent leaves that are covered with long trichomes, and glandular- or eglandular-pubescent inflorescence rachides and hypanthia. In representatives of *A. purpurascens* with strigose leaves the trichomes are less than 0.5 mm long, whereas in *A. rubricaulis* the hairs are more than 1.5 mm long.

Aciotis purpurascens may hybridize with *A. indecora*; see Appendix 2.

The illustration of *A. purpurascens* in Aublet (1775, pl. 154, as *Melastoma purpurascens*) shows strigose stems and petioles, characters found in specimens from the Guianas but not from other localities. The color plate in Bonpland (1821, pl. 48, as *Rhexia sileniflora*) shows a rather thick-walled ovary, a characteristic not easily observed in pressed specimens but very evident in fresh material.

Almeda (1997a) listed *A. purpurascens* var. *alata* Cogniaux as a synonym of *A. purpurascens* and designated *Fendler 149* (GH) as lectotype; however, Cogniaux (1885) did not propose a new variety but a new combination based on *Spennera alata* Beurling. He cited Beurling's type, *Billberg s.n.*, as well as *Fendler 149.* Because *Billberg s.n.* is now considered to belong to *A. rubricaulis*, Cogniaux's variety is a synonym of that name and not of *A. purpurascens*. That Cogniaux erred in assigning *Billberg s.n.* to *A. purpurascens* does not affect the typification of *A. purpurascens* var. *alata.*

When Stehlé et al. (1949) proposed the combination *A. purpurascens* var. *martinicensis*, they also listed *Duss 4638*, *Stehlé 5103*, and *Stehlé & Stehlé 5086*; however, these collections belong to *A. acuminifolia*.

The name *Spennera pellucida* var. *subcordata* Crueger is here considered a synonym of *A. purpurascens* on the basis of the description. The type is said to have subcordate and glabrescent leaves, which places it in A. *purpurascens* instead of *A. acuminifolia*, the only other species of *Aciotis* found in Trinidad. The leaves of A. *acuminifolia* are ovate and pubescent.

Cogniaux (1885) listed two syntypes for *A. circaeifolia* var. *major*, *Gabriel 1108* (G!) and *Spruce 337* (M! NY!). The latter is a more complete collection, and I have designated the duplicate at M as the lectotype.

During my field work I observed a population of *A. purpurascens* in eastern Ecuador in which the flowers were pollinated by halictids and also visited by dipters and *Paratrigona* Schwarz 1938 bees.

As the list of vernacular names indicates, *A. purpurascens* is well known to local people and various uses have been reported. The label of *Boom and Grillo 6356* mentions that a leaf decoction of *Tu'ñen utu' panëto' mechu' chan* is medicinal when drunk. The fruits of *Wild sauari* are eaten in Guyana (*Andel et al. 726*). Boiled leaves of *A. purpurascens* are used as cold medicine in Guyana (*Hoffman et al. 901*). Crushed leaves are used in Loreto, Peru, for children's baths (*Treacy & Alcorn 20*).

11. Aciotis rubricaulis (Martius ex DC.) Triana, Trans. Linn. Soc. London 28: 52, pl. 3, fig. 36b. 1871. *Spennera rubricaulis* Martius ex DC., Prodr. 3: 115. 1828.—TYPE: BRAZIL. Bahia: "in sylvis aeternis ad fluv. Hahype" [Rio Itaipe], Nov 1818, *Martius s.n.* (holotype: M!, photos: F! G! GH! NY!, fragment: BR!).

Arthrostema caulialatum Ruiz & Pavón, Fl. peruv. t. 327. 1830 [see note below]. *Spennera caulialata* (Ruiz & Pavón) Naudin, Ann. Sci. Nat. Bot., sér. 3, 14: 149. 1850. *Aciotis caulialata* (Ruiz & Pavón) Triana, Trans. Linn. Soc. London 28: 52. 1871.—TYPE: PERU. *Ruiz & Pavón s.n.* (holotype: MA, photos: F! NY!).

Spennera panicularis Naudin, Ann. Sci. Nat. Bot., sér. 3, 14: 148, pl. 5, fig. 1. 1850.—TYPE: PERU. Lima: "in Peruvia prope Limam," Jul 1839, *Gay s.n.* (holotype: P) [see note below].

Spennera alata Beurling, Kongl. Vetensk. Acad. Handl. 40: 125. 1854. *Aciotis purpurascens* var. *alata* (Beurling) Cogniaux in Eichler, Fl. bras. 14(3): 474. 1885. *Aciotis alata* (Beurling) Almeda, Novon 7: 335. 1997.—TYPE: PANAMA. Colón: Portobelo, "in montibus" [09°33′N, 79°37′W], 1825, *Billberg s.n.* (holotype: S!, photos: CAS, MO!).

Aciotis levyana Cogniaux in Eichler, Fl. bras. 14(3): 460. 1885.—TYPE: NICARAGUA. Chontales: Chontales, 1897, *Levy 497* (holotype: BR!; isotypes: G! US!, photos of G isotype: F! GH! NY!).

Aciotis aristata Ule, Notizbl. Königl. Bot. Gart. Berlin 6: 352. 1915.—TYPE: BOLIVIA. Pando: "Alto Acre: im Walde zwischen Porvenir und Cobija" [11°14′S, 69°05′W], Jan 1912, *Ule 9667* (holotype: B, destroyed, photos: F! GH! NY!; isotypes: G! K!).

Aciotis aristellata Markgraf, Biblioth. Bot. 29: 116. 1937.—TYPE: ECUADOR. Tungurahua: "Tal des Río Pastaza, bei Río Negro," 15 Sep 1933, *Diels 1049* (holotype: B, destroyed).

Aciotis purpurascens var. *martinensis* J. F. Macbride, Field. Mus. Nat. Hist, Bot. Ser. 13(4): 299. 1941.—TYPE: PERU. San Martín: San Roque, 1350 m, *Ll. Williams 7493* (holotype: F!; isotype: NY!).

Aciotis asplundii Wurdack, Phytologia 35: 2. 1976.—TYPE: ECUADOR. Pastaza: Shell Mera [01°05′S, 78°03′W], 900 m, 3 Apr 1956, *Asplund 20105* (holotype: S!, photos: CAS! F! NY! WIS!).

Herbs 0.3–1 m tall, erect to prostrate; stolons present; young stems quadrangular with conspicuous wings, pubescent with white to pinkish long eglandular trichomes >1.5 mm long. Leaf blades (2–) 4–7.5 (–15) cm long, (1.2–) 2.1–13 cm wide, narrowly elliptic to ovate, apex acute to acuminate, base attenuate to cuneate (rarely cordate), margin serrulate, membranous, 5–7-veined; adaxially green, pilose with long eglandular whitish trichomes, abaxially green to purplish, pubescent with long eglandular whitish trichomes, without web-like striations; petiole 1–3.5 (–4.5) cm long, with evident wings, pubescent with long eglandular whitish trichomes >1.5 mm long. Inflorescence a thyrse, 10–20 cm long, 3–8 cm wide, with (2–) 4–7 pairs of first order paraclades; rachis pilose or with glandular trichomes; internodes 2.5–4 cm long; first order hypopodia 1.2–2 cm long, biparous; second order hypopodia 0.6–1.5 cm long, biparous; third order hypopodia 0.2–0.9 cm long, biparous or uniparous; fourth order hypopodia absent to ca. 0.3 cm long, uniparous; terminal paraclade 1–2.5 cm long, with 2–4 flowers per cm; bracts 0.4–1.8 (–3) mm long, ca. 0.3 mm wide, lanceolate, apex aristate, abaxially with eglandular trichomes. Hypanthium 2–3.5 mm long, 2–3 mm in diameter, terete, without filiform sclereids in parenchyma, abaxially hirsute with glandular or eglandular trichomes, adaxially glabrous. Petals 2.7–6 (–8) mm long, 1.3–1.5 (–3.2) mm wide, triangular-ovate, apex acuminate, base truncate, with a terminal glandular trichome, abaxially glabrous. Stamens with filaments 1–2.5 mm long, anthers 1–1.8 mm long, white, pedoconnective 0.4–0.6 mm long. Ovary without filiform sclereids in parenchyma, apex glabrous. Fruit baccate, pericarp and placenta caducous when ripe, (2–) 3–5 mm long, (3–) 3.5–5 mm in diameter, bilocular, pericarp membranous, reddish green when ripe, translucent; seeds 0.60–0.70 mm long, 0.40–0.50 mm wide. Figs. 3c, d, 4b, 6b, 28.

Phenology. Flowering and fruiting throughout the year.

Distribution (Figs. 29, 30). Central America (except El Salvador) and South America (except Argentina, Chile, Paraguay, and Uruguay); disturbed, especially swampy habitats; 10–1650 m.

Vernacular names. *Chulco* (Colombia: Nariño); *pin tape* (dialect, Ecuador: Esmeraldas; Cayapa); *uchi' chinchak* (Ecuador: Morona-Santiago); *rebonilla rosada* (Peru: Pasco).

ADDITIONAL SPECIMENS EXAMINED. **Belize.** Sittee, *Peck 870* (GH).—EL CAYO: Hummingbird Highway, *Gentle 8601* (BM, CAS, F, IJ, G, LL, MEXU, MICH, MO, NY, S, TEX, UC, US), *Gentle 8249* (LL, TEX).—STANN CREEK: 17 mi Section, 16°58′N, 88°13′W, *Gentle 9248* (BM, CAS, F, G, IJ, LL, MICH, NY, S, TEX, UC, US), *Gentle 1894* (GH, MICH, NY).—TOLEDO: Punta Gorda, about 10 mi from Forestry Reserve, *Dwyer 15153* (MO). **Guatemala.** IZABAL: El Estor, 15°32′N, 89°21′W, *Contreras 11124* (CAS, LL, MEXU, MO, US, S); Puerto Barrios, *Deam 140* (F, GH, MICH, US); along banks of Río Dulce, *Muenscher 12580* (F); Cerro San Gil, *Steyermark 41855* (F); entre El Benque y La Ruidosa, en carretera hacia Petén, 15°30′N, 88°49′W, *Tún Ortiz 2438* (BM, F, MICH, US). **Honduras.** CHOLUTECA: Coyol, *Carleton 492* (US).—CORTÉS: Cuyamel, *Carleton 461* (US).—GRACIAS A DIOS: Río Plátano–Tiro Creek, NW of Bulevar, 15°43′N, 84°50′W, *Saunders 1205* (NY). **Nicaragua.** CHONTALES: Castillo El Viejo, *Shimek & Smith L. s.n.* (F); La Libertad, *Standley 9106* (F); La Libertad, *Tate 70* (BM).—MATAGALPA: Cerro Musún, al lado del Río Bilampi, 4 km SW de Wanawás, *Araquistain & Moreno 2680* (MO); Estancia El Comelar, carretera Matagalpa–Waslala, costado NE del Cerro Peñas Blancas, *Grijalva & Moreno 3463* (MO); El Comején, 1 km W de la carretera a Waslala, *Moreno & Robleto 20585* (MO); ca. 5.9 km E of Río Babasca and 2.7 km W of Río Mancera, *Stevens & Moreno 19228* (CAS, MO).—RÍO SAN JUAN: Río Indio, *Atwood 5179* (GH, MICH, MO); Mpio. El Castillo, 11°01′N, 84°24′W, *Espinosa 5187* (BM, GH, MO, NY, UC); valle del Río Indio, 6 km arriba de unión con Caño La Pimienta, *Neill 1498* (MO); near Caño Chontaleño, 20 km NE of El Castillo (Río Indio watershed), *Neill 3342* (BM, CAS, MO); San Juan del Norte, 10°55′N, 83°42′W, *Nelson 5275* (BM, GH, MO); San Juan del Norte, rivière *215* (MO); Mpio. San Juan del Norte, *Rueda et al. 2689* (MO).—ZELAYA: between Waspan and Puerto Cabezas, 14°44′N, 83°58′W, *Atwood 4818* (BM, GH, MICH, MO, NY, UC); Bluefields, *Bunting & Licht 1245* (F); Region of Braggman's Bluff, *Englesing 272* (F); 3 km N of RAMA, *Hamblett 451* (F, GH, MO, NY, UC); El Recreo, near Esperanza, *Harmon & Fuentes 5151* (MO, NY); Bahía de Bluefields, Río Escondido, 12°05′N, 83°45′W, *Molina 2070* (F,

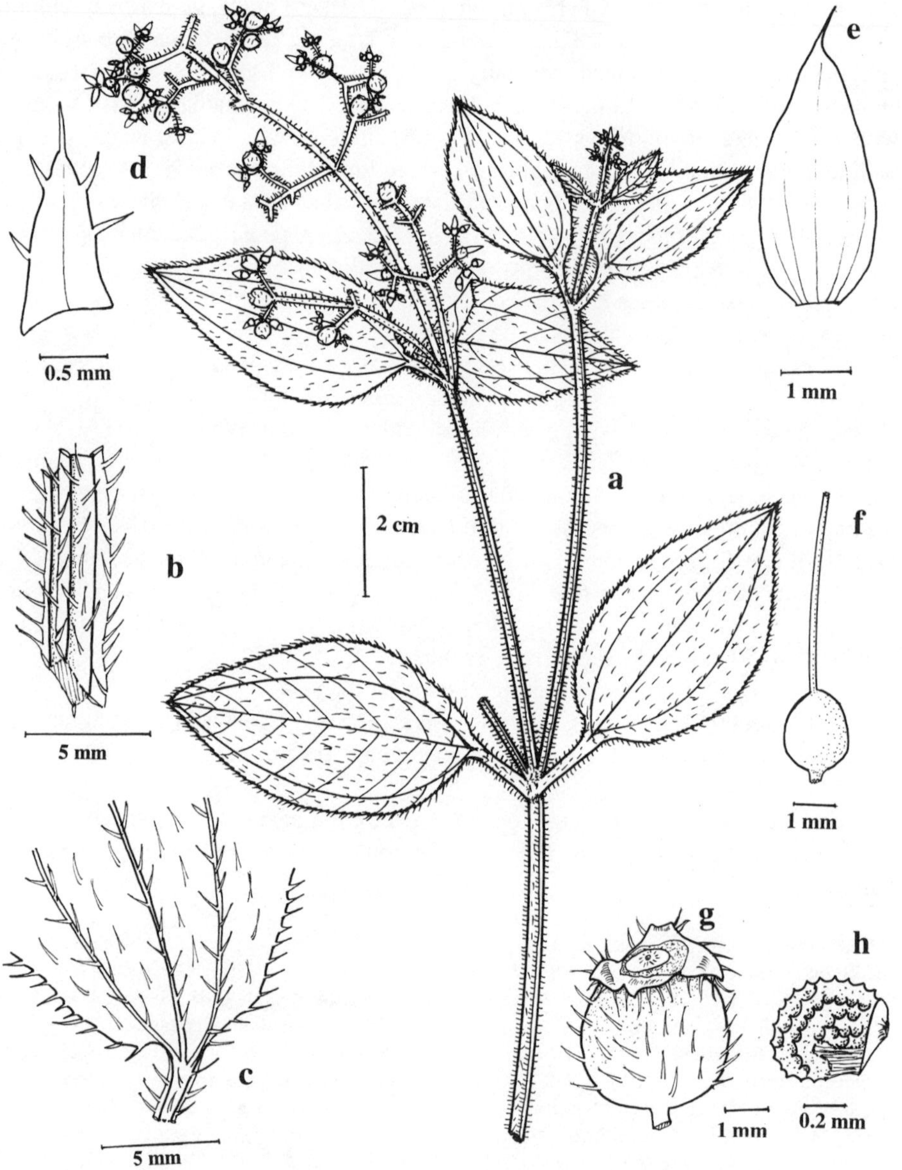

FIG. 28. *Aciotis rubricaulis*. a. Habit. b. Portion of stem. c. Abaxial leaf surface near leaf base. d. Bract. e. Abaxial surface of a petal. f. Pistil. g. Young fruit. h. Seed. (Based on: a, *Hurtado & Alvarado 1091*; b, d–h, *Freire-Fierro & León 2760*; c, *Woytowski 8182*.)

GH); Monkey Point, *Moreno & Sandino 12097* (MO); Río Punta Gorda, Atlanta, 2 km al S de Carolina del Sur, *Moreno & Sandino 12875* (MO); desembocadura del Caño el Guineo, *Moreno & Sandino 12847* (CAS, MO); desembocadura del Caño del Oro, en el Río Chiquito, *Moreno & Sandino 12961* (MO); La Richard, 11°32′N, 84°05′W, *Moreno & Sandino 13013* (MO); La Barra de Punta Gorda, *Moreno 13190* (MO); Caño Montecristo, del campamento Germán Pomares hacia arriba hasta Las Benitas, *Moreno & Sandino 14914* (MO); desembocadura del Caño El Consuelo, *Moreno & Sandino 15058* (MO); Cuesta El Bálsamo; *Moreno & Robleto 20669* (MO); Río Almacén, 8 km SW of Nueva Guinea; *Nee & Vega 27942* (MO, NY); near Bil Tingnia, 6 km NW of

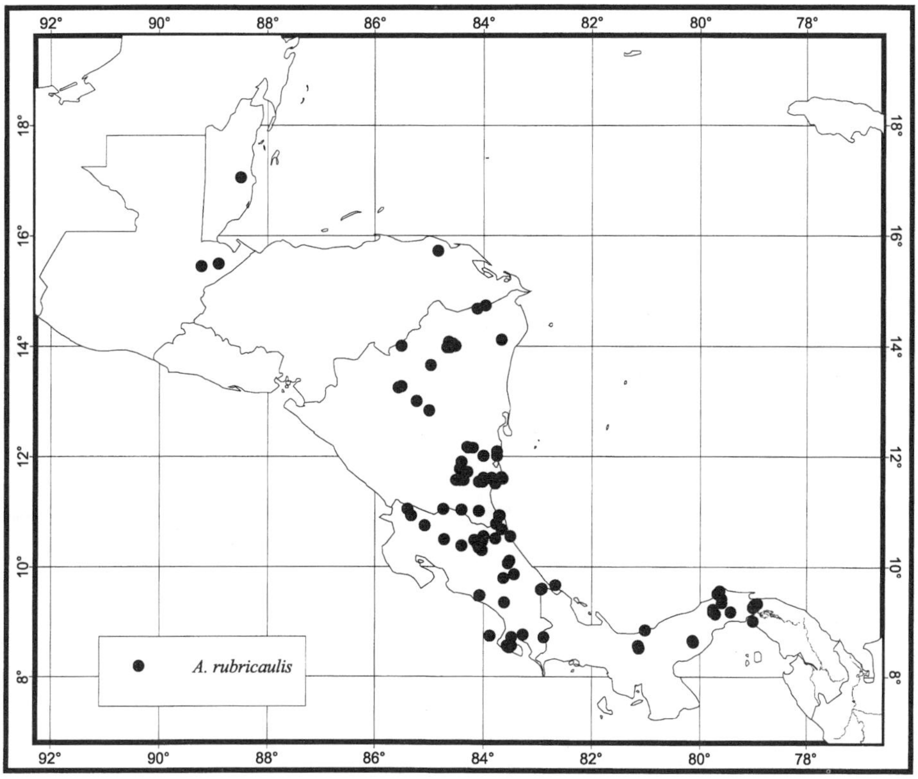

FIG. 29. Distribution of *Aciotis rubricaulis* in Central America.

Bonanza, *Neill 3978* (CAS, MO); Comarca El Cabo, Cororia Bush, 40–45 km SW of Waspan, *Seymour 3796* (GH, MO); Jackson Farm, W of Bluefields, *Seymour 4136* (BM, GH, MICH, MO, NY, UC); France ya Sirpi, *Seymour 5652* (GH); El Salto, along Río Pis Pis, *Pipoly 3538* (CAS, MO); between Kururia and San Jerónimo, *Pipoly 3769* (MO); Cerro San Isidro, *Proctor et al. 27007* (NY, US); Estación Experimental El Recreo, *Ríos 245* (MO); Nueva Atlanta, *Rueda et al. 3329* (MO); SE de Río Mico, *Sandino 1591* (CAS, MO); NW de Estación Experimental El Recreo, *Sandino 2722* (MO); 1 km de Colonia Serrano, Río Serrano, *Sandino 3437* (CAS, MO); 5 km N de colonia Jacinto Baca Jerez, *Sandino 4660* (CAS, MEXU, MO); Caño Costa Riquita, ca. 1.8 km SW of colonia Naciones Unidas, above (S of) road between Colonia Nueva León and Colonia Naciones Unidas, *Stevens et al. 5013* (BM, CAS, MO); along road from Bonanza (airstrip) through Constancia (mine shaft) to Laguna Siempreviva, 13°58′N, 84°37′W, *Stevens 8014* (CAS, MO); bridge over (Caño) Shilam Wasito, ca. 6.0 km E of first suspension bridge E of Rosita on road to Bonanza, *Stevens 8099* (MO); ca. 14.5 km W of Río Wawa, Puerto Cabezas–Rosita, at (Caño) Kauhru Tingni, *Stevens 8640* (CAS, MO); Río Blanco–Río Copalar, *Stevens 12062* (CAS, MO); Constancia–Laguna Siempreviva, *Stevens 12392* (CAS, MO); Bonanza–Constancia, *Stevens 12463* (CAS, MO); Rosita–Bonanza, bridge over (Caño) Shilam Wasito, ca. 6 km E of first suspension bridge E of Rosita on road to Bonanza, *Stevens 12525* (CAS, MO); Plantel El Salto, *Stevens 18860* (GH, MO); Bahía Monkey Point, *Stevens 20688* (CAS, MO); Cañón El Guineo, *Téllez et al. 4886* (MEXU, MO); El Diamante (La Richard), 12 km N of Atlanta, Río Punta Gorda, *Téllez et al. 5051* (MEXU). **Costa Rica.** ALAJUELA: Naranjo–Aguas Zarcas, 8.5 km NE of Villa Quezada, *Croat 46959* (CAS, MO); E slopes of Volcán Miravalle, W of Bijagua, near Río Zapote, 10°44′N, 85°05′W, *Burger et al. 11673* (BM, F, NY); Upala Dos Ríos, 5 km al Sur de Brasilia, margen derecha del Río Pizote, *Herrera 1006* (MO); Los Chiles, Río Frío, *Holm & Iltis 943* (A); 4 km SE of Fortuna, then 2.5 km SW on jeep road, *Liesner et al 15196* (MO, WIS); Artezalea, Ciudad Quezada, about 8 km NE of Villa Quezada, Quebrada Marín, *Molina et al. 17141* (F, NY); Llanura de San Carlos, near Fortuna, *Molina et al. 17634* (F, NY); San Carlos, near rivulet, *Smith 3* (F, NY).—CARTAGO: Moravia–Turrialba, *Burger & Ramírez 3962* (F, NY, US); Turrialba, Pavones–Chitaría, *Jiménez 743* (F); vicinity

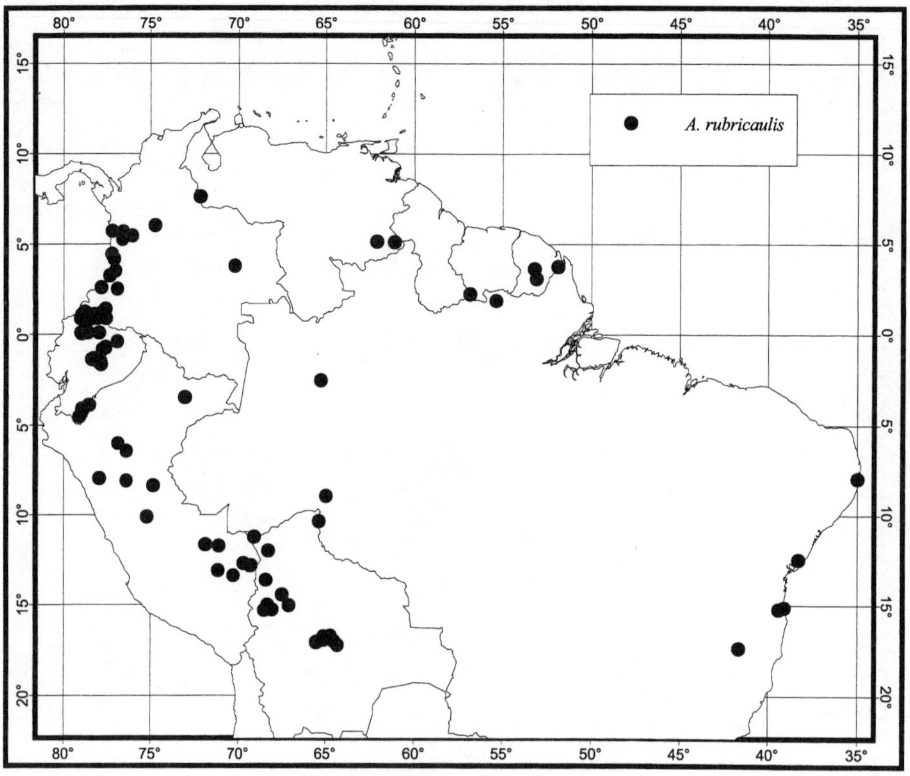

FIG. 30. Distribution of *Aciotis rubricaulis* in South America.

of Pejivalle, *Standley & Valerio 47075* (US); Río Atirro, ca. 2 km E of La Esperanza, 09°47′N, 83°38′W, *Taylor 17956* (NY); Platanillo, *Wilbur & Stone 10602* (C, DS, F, GH, LL, MICH, MO, NY, TEX, UC).—GUANACASTE: Parque Nacional Guanacaste, Estación Pitilla, 11°02′N, 85°24′W, *Chavarría et al. 23* (MO); Estación Maritza, *Delgado 98* (MO).—HEREDIA: Cariblanco, 3 mi S, road to Colonia Virgen de Socorro, *Almeda et al. 3195* (CAS); Parque Nacional Braulio Carrillo, La Virgen, Sarapiquí, *Ballesteros 35* (CR, MO); Tirimbina, *Burger & Burger 8114* (F, S); Finca La Selva, OTS Field Station, Holdridge Trail, *Folsom 9998* (CAS, MO); Río Toro Amarillo, 6 km SO de Guápiles, *Jiménez 2118* (F, NY); 1/2 km N of Puerto Viejo, 10°28′N, 84°01′W, *Mori & Anderson 52* (BM, F, WIS); Puerto Viejo, Transect F, *Opler 525* (F); Río Sucio, *Sánchez et al. 507* (AAU, F); Guápiles, *Standley 37023* (US); Estación Penal Magsasay, Sendero Principal, *Umaña & Chacón 422* (CAS); Sarapiquí, Horquetas, Estación El Plástico, Senderos El Tigre y Atajo, *Vargas et al. 1656* (F).—LIMÓN: Talamanca cantón, close to Cahuita, beyond Suretka, *Barringer et al. 2644* (F); Tortuguero, between Bri Bri and Sixaola, NW of Paraíso, 20–70 m, ca. 09°39′N, 82°40′W, *Barringer et al. 3745* (F); Reserva Indígena Talamanca, camino entre Amubri y Cachabri, del puente sobre Río Lari, margen derecha, *Chacón 3* (CR, MO); Puerto Limón, *Cufodontis 558* (G, W); Siquirres, *Davidse et al. 1179* (F); Pueblo Nuevo, 1 km SW, *Davidse & Pohl 1240* (F, MO); Hda. Tapezco–Hda. La Suerte, *Davidson & Donahue 8702* (US); Jiménez, Santa Clara, *Donnell Smith 4791* (F, G, GH, K, NY, US); Río Toro Amarillo, *Godfrey 66312* (MO); Río Santa Clara, 1.6 km E of Guápiles, *Holm & Iltis 340* (A, G, NY); Alto Lari, *Jiménez 1886* (F); 37 km NE of Turrialba on hwy to Limón, 8 km N of Linda Vista, *Liesner et al. 15448* (MO); Barra de Colorado, *Schnell 572* (F, US); drenaje de los Ríos Parismina y Reventazón, *Shank & Molina 4215* (F, GH); Río Reventazón, *Shank & Molina 4421* (F, GH); Río Colorado, Km 2 S, *Stevens et al. 25060* (CAS).—PUNTARENAS: Golfito–Television Tower, *Almeda et al. 3089* (BR, CAS); Paso Los Indios–Quepos, ca. 1–2 km E of Paso Los Indios and due NE of Quepos, *Almeda et al. 3410* (BR, CAS); Río Naranjo, near Londres and Villa Nueva, *Burger et al. 12229* (F); Los Planes (La Gloria), *Castro 278* (MO); Río Esquinas, *Cordero 198* (MO); Isla del Caño, 17 km NW of Pta. Llorona, Corcovado National Park, 08°44′N, 83°53′W, *Janzen 11643* (MO); Osa Peninsula, *Kennedy 1984* (MO); Sirena, Pavo Trail,

Kernan 164 (CAS, F, MO, TEX); Rincón de Osa, *Liesner 1868* (MO, WIS); Los Mogos, Fila Mogos, *Morales et al. 1939* (CR, NY); El Faro, Rancho Quemado, *Quesada 240* (CAS); Cantón de Osa, Agua Buena, *Ramírez 188* (MO); Osa Peninsula, 08°42′N, 83°29′W, *Raven 21602* (F, MO, NY); Rincón de Osa, *Schnell 519* (F, IJ, US); Golfo Dulce and Río Térraba, *Skutch 5275* (F, MICH, US).—SAN JOSÉ: La Hondura, 09°20′N, 83°37′W, *Standley & Valerio 51886* (US). **Panama.** BOCAS DEL TORO: Santa Catalina, *Blackwell et al. 2744* (MO); Almirante, Bomba, 09°13′N, 79°45′W, *Blum 1326* (MO); Campamento del IRHE, Changhinola 1, SE and NE, *Correa et al. 3381* (CAS, MO); Quebrada Chica, *Correa et al. 3856* (MO); N of Cerro Brujo and E of Bahia de Green, *Peterson & Annable 7197* (MO); Water Valley, vicinity of Chiriqui Lagoon, *von Wedel 1450* (GH, MO).—COCLÉ: 4 mi W of Portobello, *Antonio 1769* (CAS, MO); Río Guanche, upstream from bridge, *Antonio 4823* (CAS); N of Diamante Trial to forest ridge NW of abandoned mine on Quebrada de La Mina, 09°24′N, 79°35′W, *Churchill & de Nevers 4254* (CAS, MO); Santa Rita Trail, *Cowell 139* (NY, US); Río Turquesa, vicinity of Tyler Kittredge, *Croat 27255* (MO); vicinity of Corrosion Laboratory, *Duke 4467(a)* (UC); above El Cope, near continental divide, *Hammel 772* (CAS, MO); Llano Grande–Coclecito, Km 7 N, *Hammel 1925* (CAS, MO); Peluca, ca. 27 km from Trans-isthmian Hwy on road to Nombre de Dios, *Kennedy 2634* (MO, US); El Valle de Antón, La Mesa, *Kennedy et al. 3014* (MO); Portobelo–Río Guanche Hwy., 09°30′N, 79°40′W, *Knapp 993* (CAS, LL, MO); Río Boquerón, above Mina Boquerón, *Knapp & Sytsma 2446* (CAS, MO); Boca del Toabré and Río Coclé del Norte, *Lewis et al. 5476* (COL, MO, UC); Agua Clara, on Trinidad River, Canal Zone, *Pittier 3986* (US); El Valle de Antón–Cerro Pilón, *Porter et al. 4377* (MO, UC); Cerro Gaital, on the N-facing slope of Cerro Gaital, *Reveal & Balogh 4963* (US); Canal Zone, between France Field and Catival, *Standley 30441* (US); vicinity of Fort Sherman, *Standley 31123* (US); near Fort Randolph, *Standley 28620* (US); Pipeline Road, *Sytsma 1266* (CAS, MO); 2.5 km N of El Valle, *Sytsma & D'Arcy 3571* (CAS).—PANAMÁ: Altos de Pacora, *Antonio 3231* (CAS, MO); Campo Tres, 5 km NE of Altos Pacora, *Busey 821* (F, LL, MO, NY); Goofy Lake, 3 mi, near Cerro Azul, *Croat 11580* (F, MO); Pipeline Road, 10 to 15 km N of Gamboa, *de Nevers 3898* (CAS, MO); Cerro Azul–Cerro Jefe, *Hansen 3018* (LL, MO, WIS); El Llano–Cartí Road, 8.2 mi from Pan-american Hwy., 09°15′N, 79°00′W, *Huft & Knapp 1592* (CAS, LL, MO); Río Chagres, 5 mi SW of Cerro Brewster, *Lewis et al. 3400* (COL, MO, UC); Río Piedras, 25 km NE of Cerro Azul, *Mori & Kallunki 3476* (AAU, MO); above Río Indio, *Steyermark 17417* (MO); slopes of Cerro Jefe, beyond Cerro Azul, *Wilbur & Weaver 11351* (F, MO, US).—SAN BLAS: Nusagandi, 09°18′N, 78°58′W, *McDonagh et al. 161* (BM, MO); El Llano–Cartí Road, *Sytsma & Sytsma 3119* (CAS, LL, MO).—VERAGUAS: Río Primero Braso, 2.5 km beyond Agriculture School Alto Piedra, *Croat 25452* (MO); 12 km NW of Sante Fe, *D'Arcy 10332* (MO); Santa Fé–Río Calovébora, *Hernández et al. 750* (CAS, MO); Boca de Concepción, Golfo de los Mosquitos, 08°50′N, 81°00′W, *McPherson 11392* (MO); near Sta. María, *Witherspoon & Dressler 8917* (MO).

Colombia. AMAZONAS: Leticia, Parque Nacional Natural Amacayacu, Centro Administrativo Mata-mata 4, 03°47′N, 70°15′W, *Rudas et al. 1380* (MO).—ANTIOQUIA: San Luis, Quebrada La Cristalina, W Bh-T/Bmh-T, Sector CW, 06°00′N, 74°45′W, *Ramírez & Cárdenas 1331* (COL, HUA, MO).—CAUCA: El Tambo, La Romelia, Parque Nacional Natural Munchique, Km 85–91, *Lozano & Ruiz 7081* (COL).—CHOCÓ: Intendencia del Chocó, Río Atrato, *Archer 2036* (US); Quibdó–Medellín, Quebrada La Platina, *Arias et al. 156* (CAS, COL, MO); Río Atrato, Samurindó, *Cuatrecasas & Llano 24089* (COL, US); Vereda Partidas, *Díaz 3499* (COL); Región del Río Pichimá, Comunidad Indígena Waunana, *Forero 728* (COL); Guayabal, Río Tolo, *Forero et al. 944* (COL, MO, S, US); San José del Palmar, *Forero & Jaramillo 2430* (COL, MO); Hoya del Río San Juan, Palestina, *Forero et al. 3811* (COL, MO); Itsmina–Quibdó, entre El Dos y Las Animas, 05°15′N, 76°40′W, *Forero et al. 5440* (COL, MO, US); Tutunendo–El Carmen, Alto Río Atrato, *Forero et al. 5864* (COL); Hoya del Río Torito (afluente del Río Hábita), Finca Los Guaduales, *Forero et al. 6622* (COL, MO, US); Cerro de la Mojarra, *Forero et al. 9546* (COL, MO); Bolívar–Quibdó road, near Km 180, *Juncosa 1137a* (MO); Nuquí, Alto de Buey, *von Sneidern 36* (A, BM, COL, LL, MICH, NY, S, US).—META: Sierra La Macarena, Vereda El Tablazo, orillas del Caño Venado, afluencia del Río Güejar, *Chaparro et al. 16* (COL, F, US); Acacías, *Uribe 1305* (COL).—NARIÑO: Río Telembí, Las Mercedes, *Aguirre et al. 520* (COL); Gorgona Island, Inderena, trayecto paralelo a la costa, desde la prisión hasta la Playa de San Juan, 02°36′N, 77°54′W, *Fernández et al. 7283* (COL); Correg. Santander, Buenavista–Barbacoas (Vertiente del Río Telembi), *García-Barriga 13135* (COL, NY, US); Finca Salazar, 7 km above Ricaurte, *Gentry et al. 35016* (COL, MO); E side of Gorgona Island, *Killip & García 33133* (BM, COL, F, HUH, NY, S, US); Guapi, camino a Playa Blanca, *Lozano & Rangel 5615* (COL, MO); Barbacoas, Sitio El Barro, Reserva Los Colibríes, *Lozano et al. 6828* (COL); ca. 7 km S of Chucunez, along trail to Pialapi beyond Quebrada La Calledita, 01°10′N, 77°55′W, *Luteyn & Sylva 13940* (NY); Mpio. Piedracha, Reserva de Planadas, *Madrigal et al. 753* (COL, HUA, MO); Ricaurte–Altaquer, *Mora 3973* (COL); Pialapi–El Arrayán, *Mora 4068* (COL); Altaquer–Junín, Mpio. Altaquer, *Mora 4146* (COL); Tumaco, Km 80, *Mora 4198* (COL); Ricaurte, *von Sneidern A.580* (NY, S, US).—PUTUMAYO: Sachamates, S. Francisco de Sibundoy, Quebrada de Susunga, *Cuatrecasas 11451* (COL, F, US).—RISARALDA: Río Mistrató– Puerto de

Oro trail, Río Currumay, 05°26′N, 76°05′W, *Bernal et al. 2019* (COL, US); Mistrató–San Antonio de Chamí Rd., sitio Las Partidas, *Fernández et al. 10090* (COL).—VALLE DEL CAUCA: Río Naya; Puerto Merizalde, *Cuatrecasas 14077* (F, NY, US); costa del Pacífico, R. Yurumangui, San Antonio, 03°30′N, 77°05′W, *Cuatrecasas 16109* (F, NY, US); Río Raposo, alredores de la Estación Biológica, *Idrobo 5273* (COL); Córdoba, Dagua River, *Killip 5071* (GH, NY, US); Buenaventura, 20 km E, *Killip & García 33300* (COL, NY, US); Cali–Buenaventura, Anchicaya, *Køie 4761* (C, US); Dagua Valley, *Pittier 571* (NY, US). **Venezuela.** BOLÍVAR: carretera El Dorado–La Gran Sabana, alrededores de Km 124, *Bunting 3057* (US); Chimanta Massif, NW part of Abacapatepui, 05°13′N, 62°15′W, *Steyermark 74835* (F); El Dorado–Sta. Elena road, 05°05′N, 61°08′W, *Wurdack et al. 2848* (NY).—TÁCHIRA: Cerro las Minas, bordering Quebrada Las Minas, 18–20 km SE of Santa Ana, 07°36′N, 72°13′W, *Steyermark & Liesner 119065* (MO, US). **French Guiana.** Montagnes de la Trinité, Sommet NE, *de Granville et al. 6319* (CAY); Tabulaire –Région des Emérillos, Sommet Tabulaire, ca. 40 km SE de Saül, 03°05′N, 53°05′W, *de Granville 3615* (CAY, US); Saül, Bouefmort, *Descoings & Luu 20540* (CAY); Limonade trail, 03°37′N, 53°12′W, *Mori & Gracie 18268* (US), *Mori & Gracie 18382* (CAY, US); route de Bélizon, *Mori 18497* (CAY). **Suriname.** Tafelberg, vicinity base of Augustus Falls, *Maguire 24761* (A, F, MO, NY, RB); Sipaliwini, *Rombouts 521* (BR, F). **Ecuador.** CARCHI: Lita, *Acosta Solís 12157* (F); N of San Marcos, along trail leading to Río San Juan, *Barfod 41478* (AAU, QCA, QCNE); Maldonado–Chical, 00°54′N, 78°06′W, *Freire-Fierro 2818* (MO, QCA); Lita–Ibarra Rd, near Lita, *Laweson et al. 43981* (AAU); San Marcos de Los Coaiqueres–Gualpi Bajo, *Øllgaard et al. 57243* (AAU, K, MO, NY, QCA); trail from Rafael Quindí's Finca back toward Untal, *Wormley 1586* (MO).—EL ORO: Portovelo, 21 mi W, Moro-Moro, ca. 03°43′S, 79°37′W, *Camp E–613* (NY, US); Piñas–Santa Rosa, above El Placer, *Harling et al. 15558* (GB).—ESMERALDAS: 2–4 km SE of San Lorenzo, along railroad track, 01°15′N, 78°50′W, *Boom 2549* (NY, QCNE); Río Onzolé, upstream from San Francisco de Onzolé, 00°51′N, 79°04′W, *Holm-Nielsen et al. 25725* (AAU, CAS, NY); San Lorenzo, Río Palaví, *Hoover et al. 3797* (MO); Panadero, 5 km from the river, *Játiva & Epling 2018* (NY, UC, US,); Playa Grande, Cayapa River, *Játiva & Epling 2081* (UC); Río Cayapa, Zapallo Grande, *Kvist & Asanza 40745* (AAU, MA, NY, QCNE); Concepción, Playa Rica, 00°50′N, 78°47′W, *Mexia 8404* (BM, GB, GH, K, MO, NY, S, UC, US); Borbón, *Sparre 18031* (S, TEX, US); San Lorenzo, *Sparre 18095* (S).—IMBABURA: Parambas–Pajón, *Acosta Solís 12663* (F).—MORONA-SANTIAGO: Indanza–Limón (General Plaza), *Harling & Andersson 12773* (GB, MO, US); Limón–La Unión, Km 10–12, *Harling 26062* (F, GB, K); Zamora–Zumba (along Río Jamboe), Km 5–12, *Harling & Andersson 13895* (GB, US); Gualaquiza, Bomboiza, *Lowell & Baker 392* (F); along Río Palora, 2–5 km downstream from Arapicos, *Lugo 6038* (AAU, F, GB, NY, US); Gualaquiza, Bomboiza, *Sparre 19083* (S).—NAPO: Jondache, N of Archidona, roadside, *Asplund 8830* (S); Archidona–Sarayacu Rd, 00°50′S, 77°49′W, *Freire-Fierro & León 2760* (MO, QCA, QCNE); Coca–Curaray Rd, Km 20–30, *Harling & Andersson 11946* (AAU, GB, MO, NY, US); Hollín–Loreto Rd, Km 25, Centro Challuayacu, orilla de Río Challua, en trocha hacia zona del Guagua Sumaco, *Hurtado & Alvarado 1091* (MO, QCNE, US); Coca–Lago Agrio Rd, Proyecto Payamino (IERAC), *Sobel & Strudwick 2374* (NY).—PASTAZA: valley of upper Río Pastaza, between Río Zuñag and Cashurco, *Asplund 8546* (S); Baños–Mera, *Camp E-2381* (NY); 8 km N of Puyo *Fagerlind & Wibom 1113* (S); Mera–Río Anzu, 01°20′S, 78°10′W, *Freire-Fierro 2783* (MO, QCA); Madretierra–La Isla, *Freire-Fierro 2790* (QCA); La Isla, *Freire-Fierro 2791* (MO, QCA); Puyo–El Porvenir–Puyopungo, coll. before Cotocachi, close to river Puyo, *Freire-Fierro 2808* (QCA); Cotacocha, near to Río Puyo, *Freire-Fierro 2810* (MO, QCA); Puyopungo–El Porvenir, Km 5 N, *Lugo 4924* (AAU, CAS, GB, MO); Mera, 2 km NE of Mera, *Neill 5880* (MO, NY, QCNE, US); ca. 5 km E of Mera, road to Shell Mera, *Øllgaard et al. 35519* (AAU, CAS, MO, QCA, US); El Puyo, 01°28′S, 77°59′W, *Skutch 4520* (A, F, K, MO, NY, US); Puyo–Veracruz, Km 5–6, *Sparre 17593* (S).—PICHINCHA: Nanegal, Los Puentes, *Asplund 17278* (S); Reserva Forestal ENDESA, Carretera Quito–Pto. Quito, Km 113, 10 km al N de la carretera principal, 00°05′N, 79°02′W, *Betancourt 152* (GB); Reserva Forestal ENDESA, *Freire-Fierro 57* (AAU, F, GB, NY); Nanegalito–Nanegal rd, Km 4.5, Hda. La Esperanza, *Freire-Fierro 2767b* (MO, QCA); Los Bancos–Esmeraldas, detour to Río Silanche, Km 4, *Freire-Fierro 2768b* (MO, QCA); Reserva Forestal ENDESA, *Freire-Fierro 2834* (MO, QCA); Bancos, NE slopes of Pichincha, *Padilla 130* (AAU); Mindo, Nanegal, *Sodiro 478* (BR); Nanegalito, *Sparre 14892* (GB, RB, S).—TUNGURAHUA: Old Canelos trail, below Río Margaritas, *Penland & Summers 225* (F, NY); Río Negro, 01°24′S, 78°13′W, *Sparre 17662* (S, TEX, US).—ZAMORA-CHINCHIPE: Valladolid, track NW of village, *Jørgensen 56427* (AAU); Romerillo, trail at limit of Parque Nacional Podocarpus, *Madsen & Knudsen 86830* (AAU, MO, QCA, QCNE); Quebrada del León, affluent Río Bombuscara, S. of Zamora, 04°07′S, 78°58′W, *Øllgaard et al. 90356* (AAU, MO, QCA, QCNE); ca. 4 km E of Paquisha, *Øllgaard et al. 90456B* (AAU, MO, QCA). **Peru.** AMAZONAS: Mendoza, 06°20′S, 77°24′W, *Woytkowski 8182* (MO).—CUSCO: Río Araza, N.E. of Cuzco, *Sandeman 3666* (K); Quispicanatis, Mándor–Marcapata, *C. Vargas C. 3786* (MO); Paucartambo, Atalaya, Kosñipata, *C. Vargas C. 13982* (US); Pilcopata, *C. Vargas C. 116661* (MO); Wajyumbe, *C. Vargas C. 14571* (US).—HUÁNUCO: road to Aucayacu, ca. 3 km from jct with route 16, 15 km N of Tingo Maria, *Davidson 3438* (F, MO, US); Tingo María,

Monzón road to Puñulla, ca. 09°18′S, 75°59′W, *Mathias & Taylor 5425* (F, US); Pachitea, Honoria, Bosque Nacional de Iparia, Caserío de Miel de Abejas, *Schunke 1876* (COL, F); Quebrada de Aucayacu, *Schunke 6553* (F, MO, US); Leoncio Prado, Rupa Rupa, Jacintillo Road, Río Monzón, *Schunke 8321* (AAU, F, MO, NY, US).—LORETO: Maynas, Río Afayacu, *Ayala et al. 2906* (MO, NY), *Ayala et al. 2924* (MO, NY, TEX, US); entre Pucallpa y Tingo María, *Ferreyra 17572* (CAS, US); Coronel Portillo, Neshuya, Km 64 cerca de Neshuya, *Ferreyra 18174* (US); Maynas, Bosque Nacional von Humboldt, Km 86 Pucallpa–Tingo Maria road, *Gentry & Daly 18694* (MO); Río Yavari, Emilia, above Atalaia del Norte, *Gentry & Revilla 20769* (MO, U); Coronel Portillo, Lower Boquerón del Padre Abad, *Hutchinson et al. 6049* (UC); Gamitanacocha, Río Mazán, *Schunke 213* (F, GH, NY, UC, US); Boquerón Padre Abad, *Woytkowski 34412* (F, G, MO, UC, US).—MADRE DE DIOS: Tambopata Nature Reserve, ca. 30 km air or 70–80 km river SSW Puerto Maldonado at effluence Río La Torre (Río D'Orbigny)/Río Tambopata (SE bank), *Barbour 4928* (F, MO, US); Shintuya, *Foster et al. 3107* (F); Río Manu, Río Sotileja, 11°40′S, 71°55′W, *Foster et al. 11588* (AAU, F); Manu, Manu National Park, Cocha Cashu uplands, *Núñez 6022* (MO, TEX, US); Río Salvación, *Núñez 6582* (MO); Explorer's Inn, 5 km down Main Trail, near the confluence of Río Tambopata and Río La Torre, 39 km SE of Puerto Maldonado, *Smith et al. 1009* (NY, US).—PASCO: Pichis Trail, Porvenir, *Killip & Smith 25892* (NY); Oxapampa, Palcazu, Iscozacin above PEPP project camp, 10°09′S, 75°12′W, *Salick & Quijano 7659* (NY, US).—PUNO: St. Gaban [13°24′S, 70°18′W], 1853, *Lechler 2404* (G, K).—SAN MARTÍN: Moyobamba, *Mathews 1279* (E, K); Depto. San Martín, Prov. Mariscal Cáceres, Dtto. Tocache Nuevo, Quebrada de Cascarilla, NE of Puerto Pizana, 08°09′S, 76°26′W, *Schunke 6555* (MO), *Schunke 7854* (GH, MO, XAL); Zepelacio, *Klug 3319* (F, G, GH, HUH, MO, NY, S, US).—UCAYALI: Coronel Portillo, Bosque Von Humboldt, 03°15′S, 76°36′W, *Young 991* (MO, TEX, US). **Bolivia.** COCHABAMBA: Villa Tunari–El Beni Rd, Km 40, por Chipiriri, *Beck 13728(A)* (US); Antahuacana, Espíritu Santo, *Buchtien 2909* (BR, GH, US); Puerto Villareoe, Región del Chapare, vereda de R. Murillo, *Cadima 51* (LPB, US); Puerto Polonia, Río Coni, 14 km E of San Antonio, *Cárdenas & Cutler 7340* (F, NY, US); Chapare, near Río Mateo, near Chimore, *Eyerdam 24750* (F, K, UC); José Carrasco Torrico, Cochabamba–Villa Tunari old Rd, Km 143, *Kessler et al. 8233* (MO); Todos Santos, 16°48′S, 65°08′W, *Steinbach 458* (F, HUH, MICH, MO, NY, S, US).—EL BENI: Rurrenabaque, 15°04′S, 67°07′W, *Cárdenas 1898* (MO); General Ballivián, SW of Km 12 Yucumo–Rurrenabaque, 12 km por el camino maderero, 15°04′S, 67°07′W, *Kessler et al. 10784* (MO); Cachuela Esperanza, 10°32′S, 65°38′W, *White 2092* (GH, NY, US).—LA PAZ: La Paz–Beni, Sud Yungas, Río Quiquibey, *Beck 8071* (US); San Carlos, *Buchtien 1062* (BM, C, E, G, K, S, WIS, Z); Mapiri, Charopampal, *Buchtien 1719* (NY, US); J. Bautista Saavedra M., Pauji–Yuyo, entre Apolo y Charasani, *Kessler et al. 9830* (MO); Larecaja, Tuiri (near Mapiri), on left bank of Río Mapiri, *Krukoff 10786* (NY); Mapiri, 15°15′S, 68°10′W, *Rusby 2333* (BM, E, F, G, GH, HUH, K, MICH, MO, NY, US, W); Prov. Larecaja, 6 km N (below) Consata, *Solomon et al. 6584* (MO, US); Abel Iturralde, Río Beni, Encañada Suse, *Solomon 16927* (NY).—PANDO: Manuripi, along Río Madre de Dios, 3 km W of Humaitá, 12°01′ S, 68°18′W, *Nee 31659* (NY, US).—SANTA CRUZ: Ichilo, Santa Cruz 130 km NW, puente Río Yapacani, *Beck 6564* (US); Yapacanichilo, *Steinbach 7573* (BM, F, G, GH, K, MO, S, Z). **Brazil.** ACRE: Tarauacá, Rio Muru, 12 km above confluence with Rio Tarauacá, *Prance et al. 7425* (K, NY).—AMAPÁ: Rio Oiapoque, 03°45′N, 51°52′W, *Black 49-8405* (NY); Serra do Navio, area industrial, *Cowan 38445* (NY, RB, US).—AMAZONAS: Nova Prainha, RADAM/BRASIL SB-20-ZB-Pto. 19, *Mota & Monteiro s.n.* (INPA); Rio Juruá, 02°37′S, 65°44′W, *Ule 5663* (G, K).—BAHIA: Itabuna, Camacã–Canaivieira, Km 3–30, *Belém & Aguiar 1394* (US); Ilhéus, *Blanchet s.n.* (G); Ba. Mascote, 3–4 km de estrada a Canavieiras, na margem do Rio Pardo, *dos Santos 838* (US); Itabuna, Belmonte, *dos Santos 1127* (RB, US); São Acuto dos Lagos, *Luetzelburg 204* (M); Maracas, *Martius 1044* (G); Mpio. Una, entrada no km 17 da Rod. BA-101 (São José) a Rod. BA-215, 15°09′S, 39°05′W, *Mori & Thompson 11035* (NY, RB); Caraiba Metais, Herb. Univ. Feira de Santana, *Noblick et al. 2405* (RB); Jaguaquara–Apuerema Rd, Km 36, *Pinheiro 1846* (RB, US).—PARÁ: Parque Indigena do Tumucumaque, Rio Paru de Oeste, Missão Tiriyo, arredores da Missão, 02°20′N, 55°45′W, *Cavalcante 2445* (NY, US); Acará, Thomé Assú, Agua Branca, *Mexia 6039a* (GH, K, NY).—PERNAMBUCO: Catuca, *Gardner 1007* (BM, K); Igarassu, Usina São José, *Lima 56-2608* (HB); Dois Irmãos, 08°50′N, 35°57′W, *Riddley s.n.* (BM).—RONDÔNIA: Rio Acre, Antimary, *Huber MG-4259* (HB); Track between Mutumparana and Rio Madeira, Basin of Rio Madeira, 09°N, 65°W, *Prance et al. 5556* (INPA, K, US).

Aciotis rubricaulis is easily recognized by the pubescence that covers all vegetative parts, evident wings, membranous leaves, thyrsoid to paniculate inflorescences, and baccate fruits. It has predominately thyrsoid inflorescences with four to five pairs of first-order paraclades. The Colombian collection *Mora 4146* is exceptional in having some branches with inflorescences bearing two pairs of paraclades but otherwise agrees in all the other morphological characteristics.

Aciotis rubricaulis may hybridize with *A. indecora* and *A. polystachya*; see Appendix 2.

This species has been known as *Aciotis caulialata* and *A. levyana*, but the name *Aciotis rubricaulis* has priority. Ruiz and Pavón described *Arthrostema caulialatum* as part of their *Flora peruviana, et chilensis* (1798–1802), but in volume 4, which remained unpublished until 1954–1957; however, the name was effectively published, because copies of the plate, t. 327, which includes structural details, were circulated by 1830 (Stafleu & Cowan 1983; Greuter 2001). This plate shows the dense pubescence of the vegetative parts and the conical inflorescences with up to seven pairs of paraclades (Anal. Inst. Bot. Cavanilles 12: 178. 1954).

The type collection of *Aciotis levyana* (*Levy 497*) has sepals up to 1 mm long, one of the two characters used by Wurdack (1980) to segregate some collections from eastern Ecuador as *A. asplundii*. The second character used to distinguish *A. asplundii*, cordate leaf bases, is highly variable.

The collection *Ule 9667* from Bolivia was described as *Aciotis aristata* because of the aristate petals, but falls within the variation of *A. rubricaulis*. The isotype deposited at G has petals up to 5 mm long, acuminate buds, pedoconnectives 1 mm long, and blackish leaves, much like the Colombian specimen *Cuatrecasas 14077* (NY), which also has long petals and long pedoconnectives. Although the blackish color of the leaves is not a common characteristic, it has been observed also in at least one Nicaraguan collection (*Stevens 18860*). In general, the Central American and northern South American collections have smaller leaves with cuneate bases and smaller petals, whereas the Peruvian and Bolivian collections have larger leaves with obtuse to cordate bases and larger petals.

In the protologue *Spennera panicularis* is said to be "hirto-furfuracea," and the original plate shows the leaves, petals, and stamens typical of *Aciotis rubricaulis*. This species does not occur in Peru, and the purported type locality is most likely erroneous.

The type of *Aciotis aristellata* is no longer extant, and the name is here placed in synonymy on the basis of the description. Additional representatives of *A. rubricaulis* have been collected a number of times in the area of the type locality.

Occasional dispersal by ants may occur when leaf-cutter ants cut an entire plant of *A. rubricaulis* and carry it, including the fruits, to their nests (pers. obs.)

Whiffin and Tomb (1972) examined seeds of the Costa Rican collection *Shrank & Molina 4421*, deposited at US, for a study of seeds of Neotropical capsular-fruited species of Melastomataceae.

An Ecuadorian population of *A. rubricaulis* was visited by colletids and by chrysomelids that were eating the petals (pers. obs).

In Colombia, "*chulco*" is used to cure kidney disease; in Ecuador "*pin tape*" (*Kvist & Asanza 40745*) is used for curing snake bites.

12. Aciotis indecora (Bonpland) Triana, Trans. Linn. Soc. London 28: 52. 1871. *Rhexia indecora* Bonpland, Monogr. Melast. 2: 131, pl. 49. 1821. *Spennera indecora* (Bonpland) DC., Prodr. 3: 116. 1828.—TYPE: VENEZUELA. Amazonas: "Brasilia," *Bonpland s.n.* (holotype: P, photos: F! NY! US!, fragment: F!).

Spennera laxa Richard ex DC., Prodr. 3: 116. 1828. *Aciotis laxa* (Richard ex DC.) Cogniaux in Eichler, Fl. bras. 14(3): 476, pl. 108. 1885.—TYPE: FRENCH GUIANA. Without locality, *Richard s.n.* (holotype: P, photos: F! NY!).

Spennera acutiflora Martius, Flora 22, Beiblatt 1(4): 58. 1839. *Aciotis acutiflora* (Martius) Triana, Trans. Linn. Soc. London 28: 52. 1871.—TYPE: BRAZIL.

Bahia: Ilhéus, *Martius 400* (holotype: M!; isotypes: BM! BR [3 sheets]! F! G [4 sheets]! GH! K! MO! NY [4 sheets]! US! W!).

Spennera rostellata Naudin, Ann. Sci. Nat. Bot., sér. 3, 14: 143. 1850. *Aciotis rostellata* (Naudin) Triana, Trans. Linn. Soc. London 28: 51. 1871.—TYPE: MEXICO. Tabasco: "in uliginosis prope Teapa" [18°35′N, 92°56′W], Nov 1839, *Linden 637* (holotype: P; isotypes: BR [3 sheets]! G! K!, photos of G isotype: F! GH! MICH! NY!).

Spennera viscosa Naudin, Ann. Sci. Nat. Bot., sér. 3, 14: 146. 1850. *Aciotis viscosa* (Naudin) Triana, Trans. Linn. Soc. London 28: 52. 1871.—TYPE: VENEZUELA. Amazonas: "prope Maypures" [=05°10′N, 67°47′W], *Bonpland s.n.* (holotype: P, photos: F! NY!, fragment: F!).

Spennera kappleriana Naudin, Ann. Sci. Nat. Bot. 3, 14(2): 146. 1850. *Aciotis laxa* var. *kappleriana* (Naudin) Cogniaux in Eichler, Fl. bras. 14(3): 477. 1885.—TYPE: SURINAME. "in paludosis umbrosis," 1842, *Hostmann & Kappler 1016* (holotype: P; isotypes: BM! BR! G [2 sheets]! K! MO! S! W! Z!).

Aciotis indecora var. *macrophylla* Cogniaux in Eichler, Fl. bras. 14(3): 470. 1885.—TYPE: BRAZIL. Amazonas: "in paludibus ad Ega prov. Alto Amazônas" [=Tefé, 03°24′S, 64°45′W], 1831, *Poeppig s.n.* (lectotype, here designated: W!).

Aciotis indecora var. *sagotiana* Cogniaux in Eichler, Fl. bras. 14(3): 470. 1885.—TYPE: FRENCH GUIANA. Basin de la Mana, "Acarouany," [05°30′N, 54°02′W], 1855, *Sagot 212* (holotype: BR!; isotypes: BM! BR! G! K! S! U, W!, digital photo of U isotype: MO!).

Aciotis acutiflora var. *parvifolia* Cogniaux in Eichler, Fl. bras. 14(3): 476. 1885.—TYPE: BRAZIL. Bahia: *Blanchet 208* (holotype: BR; isotype: W!).

Aciotis laxa var. *robusta* Cogniaux in Eichler, Fl. bras. 14(3): 477, pl. 108. 1885.—TYPE: BRAZIL. Bahia: "inter Victoria et Bahia," *Sello s.n.* (lectotype, here designated: BR!; isolectotype: K!).

Leandra montana Gleason, Rec. Trav. Bot. Néerl. 32: 210. 1935.—TYPE: SURINAME. Wilhemina Mts, *B.W. 7209* (*Stahel Wilh. Exp. 494)* (holotype: U, digital photo: MO!).

Aciotis anomala Brade, Inst. Nac. Pesq. Amazônica Publ. Bot. 8: 8. 1958.—TYPE: BRAZIL. Amazonas: Manaus, Campo Sales, "B. A. M. Km 24 da rd BR-17" [03°08′S, 59°00′W], 5 Jul 1955, *Francisco s.n. =INPA 1318* (holotype: INPA!; isotype: INPA!).

Herbs to subshrubs 0.5–2 m tall, erect; stolons absent; young stems quadrangular with inconspicuous wings, glandular-pubescent with trichomes <1 mm long. Leaf blades 4.5–14 cm long, 1.7–8.5 cm wide, ovate, apex acute to acuminate, base cuneate to cordate, margin serrulate, papyraceous, 5–7-veined; adaxially dark green, pubescent, with glandular and eglandular trichomes, abaxially light green, pubescent, with glandular and eglandular trichomes, without web-like striations; petiole 0.8–4.5 (–8) cm long, with narrow wings, densely glandular-pubescent, the trichomes <1 mm long. Inflorescence a thyrse or panicle, ca. (3.5–) 4.5–8 long, 6–18 cm wide, with (3–) 4–6 (8) pairs of first order paraclades; rachis densely glandular-pubescent; internodes 1.8–2.8 cm long; first order hypopodia 1.3–2 cm long, biparous; second order hypopodia 0.5–0.7 (–1.5) cm long, biparous; third order hypopodia 0.2–1 cm long, biparous or uniparous; fourth order hypopodia 0.3–0.4 cm long, uniparous (rarely biparous); terminal paraclade 1–3 cm long, with 1–2 flowers per cm; bracts 0.6–1 mm long, ca. 0.3 mm wide, ovate, the apex acute, abaxially glandular-pubescent. Hypanthium 1.8–2.5 mm long, 1–1.2 mm in diameter, terete, without filiform sclereids in parenchyma, abaxially densely glandular-pubescent, adaxially glabrous. Petals 1.5–3 mm long, 1–1.5 mm wide, oblong, apex acute, base cuneate, with a terminal glandular trichome, abaxially with glandular trichomes (rarely glabrous). Stamens with filaments 1.2–1.6 mm long, anthers (0.6–) 1–1.4 mm long, white or purple, pedoconnective 0.2–0.6 mm long. Ovary without filiform sclereids in parenchyma, apex glandular-pubescent (rarely glabrous). Fruit baccate, pericarp and placenta caducous when ripe, 2–3.5 (–5) mm long, 2.5–3 mm in diameter, bilocular, pericarp membranous, green when ripe, translucent; seeds 0.40–0.50 mm long, 0.30–0.40 mm wide. Fig. 31.

Phenology. Flowering and fruiting throughout the year.

Distribution (Figs. 32, 33). Central America (except El Salvador) and South America (except Chile, Paraguay, and Uruguay); primary and secondary forests; 10–1700 m.

Vernacular names. *Wandering Jew* (Nicaragua: Zelaya); *oluka burwid* (Panama: San Blas; Kuna); *peludita* (Panama: Panamá); *uzillabingoi* (Colombia: Amazonas; Witoto); *yuyito* (Colombia: Antioquia); *caña agria* (Colombia: Cundinamarca); *mortiña* (Colombia: Chocó); *barba de paca* and *buxixú* (Brazil: Amazonas); *boyoyo blanco* (Venezuela: Amazonas); *pio-yek* (Venezuela: Bolívar).

ADDITIONAL SPECIMENS EXAMINED. **Mexico.** CHIAPAS: Ocosingo, Ejido Boca de Chajul, 17°04′N, 92°15′W, *Domínguez 116* (US).—TABASCO: Huimanguillo, *Barlow 30/86-B* (MEXU); Huimanguillo, *Cowan 3328* (CAS, MEXU, NY, XAL); Huimanguillo, *Curiel & Rodríguez 155* (XAL); carretera Huimanguillo–Francisco Rueda, desviación a Tecominoacán, 17°51′N, 93°23′W, *Guizar 507* (CAS); near Teapa, monte Cocorrá, *Rovirosa 888* (K); without locality, *Zamudio et al. 92* (CAS, MEXU).—VERACRUZ: Rancho Gavilán, 3 km before Las Choapas, *Calzada 5971* (F, XAL); Trans-Isthimian highway, Km 4 NW of Minatitlán, *King 1116* (MICH, US); Coatzacoalcos, 6.5 mi E, *Kral 25400* (MICH, US); Jaltipán, 17°58′N, 94°42′W, *Orcutt 6514* (DS, MICH, NY, TEX, US); 2 km del Río Tonalá rumbo a Coatzacoalcos, *Orozco 147* (MEXU).

Belize. STANN CREEK: Cockscomb Basin, Jaguar Preserve, 10 km W of Maya Center, off Southern Highway, Gibnut Loop Trail, 16°45′N, 88°35′W, *Balick et al. 2607* (MO, US); Hummingbird Hwy, along southern hwy ca. 3 mi S of junction with Hummingbird hwy, 6 min. W of Stann Creek, *Croat 24215* (MO); Coxcomb Mt. Basin, E of Mts flat lands towards sea, *Dwyer et al. 476* (MO, NY); Stann Creek–Mullin's River Rd, *Gentle 1916* (A, F, K, MICH, NY, US); Carasow Hill, *Gentle 7907* (CAS, F, GH, IJ, LL, MICH, MO, S); Mullins river road, *Schipp 46* (BM, GH, K, MICH, MO, S, UC, US, Z).—TOLEDO: Honey Creek, Monkey River, *Gentle 4462* (CAS, LL, MICH, TEX); between Condemned Branch Pine Ridge and Moffredye Lagoon, *Gentle 6063* (F, LL [3 sheets], S, TEX); Chique Sarco, *Hedger 249* (BM). **Guatemala.** CHIQUIMULA: E of Vera Cruz and Chiquimula, *Watson 392a* (BR, GH).—IZABAL: Los Amates, Quebradas, *Blake 7509* (US); La Cumbre, Km 158 on Cadenas Rd, *Contreras 8952* (CAS, LL, MEXU, MO, S); Los Amates, *Deam 141* (GH, MICH, NY); W of Mariscos, Finca Carolina, 15°15′N, 89°00′W, *Haynes 7056* (NY); vicinity Lago Izabal, *Jones et al. 3044* (F, IJ, LL, MEXU, MICH, NY, UC, US); between Los Amates and Izabal, *Kellerman 7363* (F, MEXU, US); Cadenas, on old brecha to Río Chocón Arriba, *Lundell & Contreras 19013* (CAS, LL [3 sheets], MO); trail Salvador–Cálix, Biotopo Chocón Machacas, *Martínez et al. 23156* (MEXU); Pto. Barrios, 3 km S of Punta de Palma, *Martínez et al. 23626* (CAS, MEXU); Livingston, *Tuerckheim 1218* (US); near Entre Ríos, *Standley 72663* (F); 4–5 km W of Santo Tomás de Castilla, *Stevens et al. 25615* (CAS, MO, RB); Montaña del Mico, between Bananera and La Prensa, *Steyermark 38093* (F, NY); Puerto Barrios–Santo Tomás Rd, *Steyermark 42043* (F, NY).—PÉTEN: Potún, camino para Puerto Méndez, a Km 163 appr. 50 m del camino, SW side, *Tún 2073* (F, MO, NY). **Honduras.** CORTÉS: Cuyamel, *Carleton 487* (F, US).—GRACIAS A DIOS: Ahuas, La Mosquita, 15°30′N, 84°20′W, *Clewell 3649* (MO); Rus-Rus, vaguada del Río Rus-Rus, *Nelson & Romero 4099* (CAS, MO). **Nicaragua.** RÍO SAN JUAN: Santa Marta, 30 km W of San Juan del Norte, 14°18′N, 83°54′W, *Martínez & Rivière 2063* (MEXU); Boca Machado (Valle de S. Juan), *Pittier & Durand 9657* (G); entre el Pueblo de San Juan del Norte Nuevo y La casa de Ramón Castillo, viajando por el caño San Juanillo, *Rueda et al. 1853* (MO); Río Pigibaye, *Rueda et al. 3191* (MO); Río Indio, hours upriver from S. Juan del Norte, from Finca Gurie 0.5 km E through swamp and on to high ground, *Rivière 253* (MO); 1.5 km W of Delta, Río San Juan, *Rivière 321* (CAS, MO).—ZELAYA: frecuente a orillas de Río Licus, La Tronquera, Comarca El Cabo, *Molina 14949* (BM,

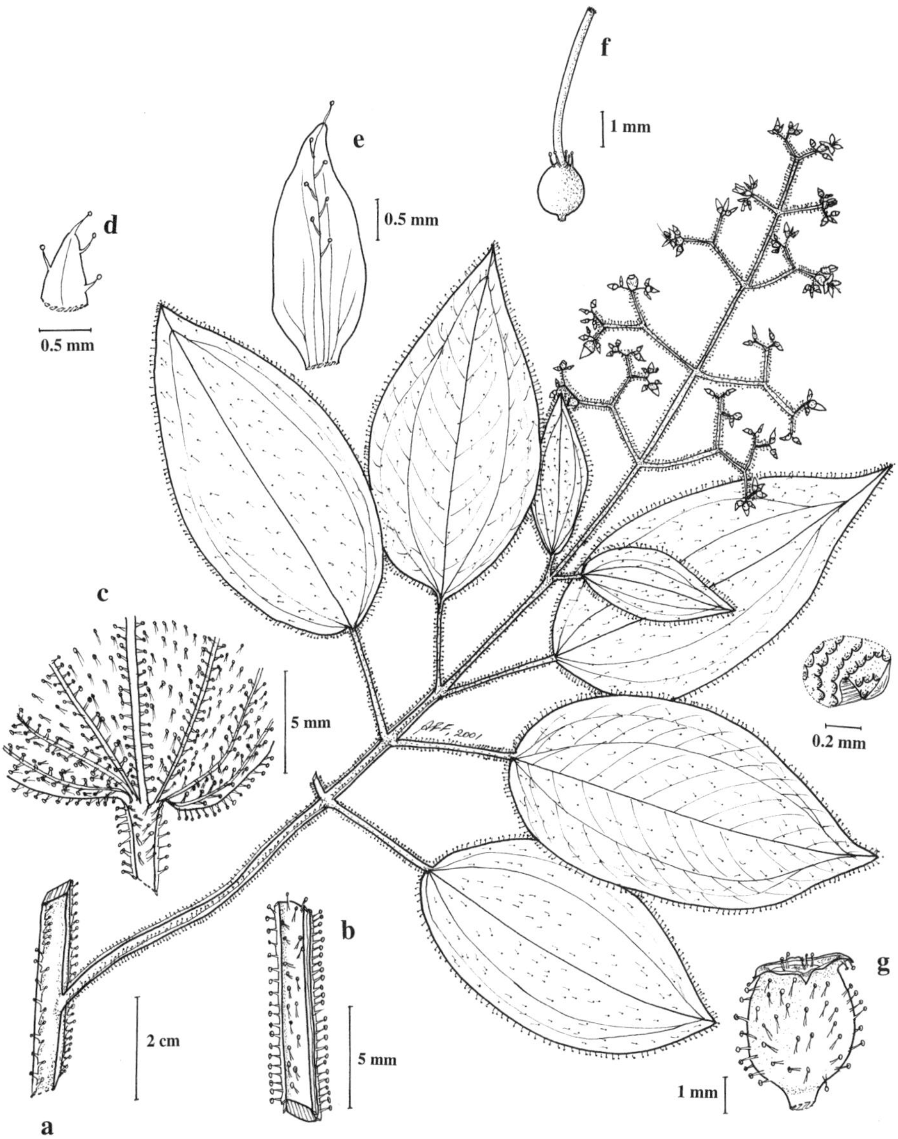

FIG. 31. *Aciotis indecora*. a. Habit. b. Portion of stem. c. Abaxial leaf surface near leaf base. d. Bract. e. Abaxial surface of a petal. f. Pistil. g. Fruit. h. Seed. (Based on: a, b, *Solomon 7592*; c, *Plowman et al. 8062*; d–h, *Almeda 7595*.)

F, GH, NY, US, W); Comarca de El Cabo, Quebrada Cuyu, *Molina 15028* (F, NY, US); Caño Monte Cristo, La Grupera, 11°33′N, 83°48′W, *Moreno & Sandino 14675* (CAS, MO); Colonia Kururia, *Pipoly 3917* (CAS, MO); 3.2 km W of Río Wawa ferry, *Stevens et al. 19562* (CAS, MO); Santa Marta, *Stevens & Moreno 19626* (CAS, MO); SW of Bluefields, *Stevens 19822* (CAS, MO); vicinity of Puerto Cabezas, *van Stelle 51* (MO). **Costa Rica.** ALAJUELA: Bijagua–Upala, 2–4 km N, *Almeda & Nakai 4031* (CAS); Finca El Roble, Los Chiles, (Llano Medio Queso), *Gómez-Laurito 10891* (F); San Carlos, Boca Tapada, 7 km al norte, camino al Río San Juan, 11°25′N, 84°10′W, *Morales et al. 1545* (CR).—HEREDIA: La Virgen–Puerto Viejo, 2.6 km N of La Virgen, *Almeda et al. 3170* (CAS); 1.4 km NW of Puerto Viejo, 10°28′N, 84°10′W, *Anderson & Mori 32* (BM, F, WIS);

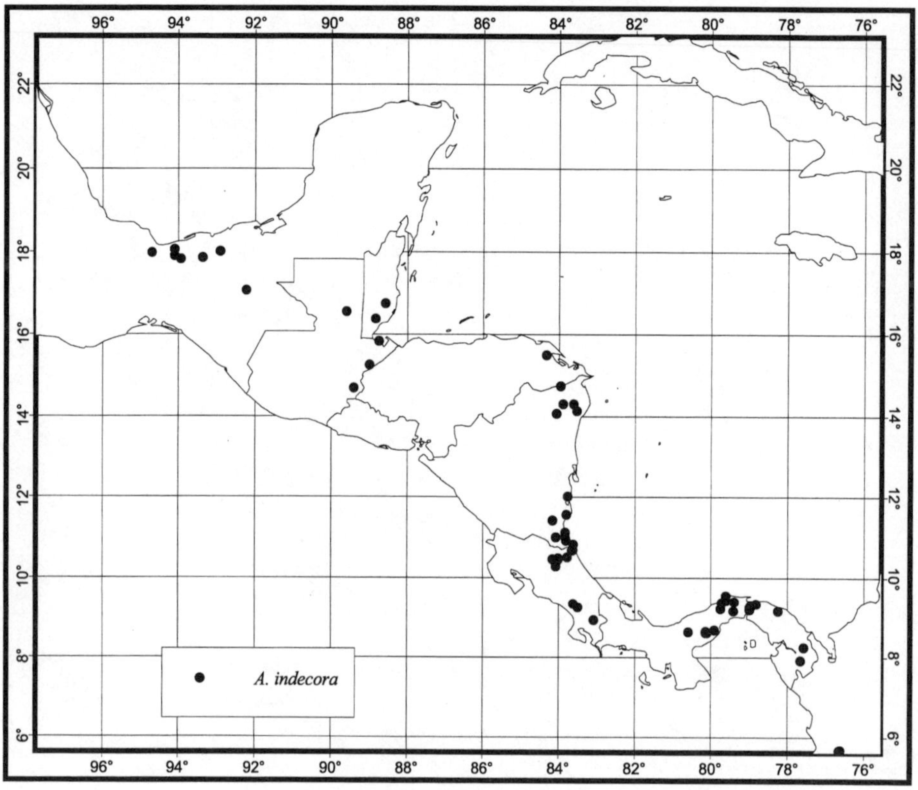

FIG. 32. Distribution of *Aciotis indecora* in Mexico and Central America.

Parque Nacional Braulio Carrillo, 200 m E del Puesto La Ceiba, sobre la Fila y rivera del Río Peje, *Ballesteros 13* (MO); El Muelle, *Biolley 7449* (BR, US, Z); Finca El Bejuco, S base of Cerros Sardinal, Chilamate de Sarapiquí, 10°27′N, 84°04′W, *Grayum & Jacobs 5347* (MO); forest between Río Peje and Río Sardinalito, atlantic slopes of Volcán Barba, *Grayum 6669* (CAS, MO); Finca La Selva, OTS Field station, *Hammel 8159* (F, MO); Finca Don Bosco, *Jiménez 4160* (F); Chilamate, *León 1774* (US); Puerto Viejo, *Schnell 306* (F, GH); Finca La Virgen de Sarapiquí, *Soleau ACM 62* (F).—LIMÓN: Hda. Tapezco–Hda. La Suerte, 29 km W of Tortuguero, *Davidson & Donahue 8375* (MO, US); Barra de Colorado, 10°40′N, 83°40′W, *Schnell 590* (F, US); E of Cerro Coronel, opposite Laguna Danto, *Stevens & Montiel 24663* (CAS); Laguna Pereira, S shores, *Stevens et al. 25174* (CAS).—PUNTARENAS: Paso Los Indios–Quepos, *Almeda et al. 3408* (CAS); San Isidro del General, *Chrysler & Roever 5283* (NY); Cantón Buenos Aires, Valle de El General, Río Convento, 09°15′N, 83°30′W, *Hammel et al. 19180* (CAS, CR, MO); Río Ceibo, near Buenos Aires, *Pittier 6516* (BR) Valle de El General, *Skutch 3892* (A, MO, NY, S, US).—SAN JOSÉ: ca.1–3 km, SW of Quizarra, 09°20′N, 83°37′W, *Almeda & Nakai 4143* (CAS); Finca El Quizarrá, Río Hermoso, lower slopes of Cordillera de Talamanca, *Williams et al. 28461* (F, NY, US). **Panama.** BOCAS DEL TORO: Santa Catalina, *Blackwell et al. 2735* (COL, MO, UC); Nances Cay Island, *von Wedel 582* (GH, MO).—COCLÉ: 8.5 km beyond the Market Place in El Valle de Antón on the road to trail head to Cerro Caracoral, 08°37′N, 80°08′W, *Almeda et al. 7595* (CAS); El Cope, *Antonio 1199* (MO); El Valle–La Mesa road, 2 mi above El Valle, *Croat 13318* (MO); La Mesa, above El Valle de Antón, N of Cerro Gaital, *de Nevers et al. 3522* (CAS, MO); El Valle–Cerro Pilón, *Dwyer 7954 A* (MO); Alto Calvario, *Folsom & Robinson 2391* (CAS); La Mesa, N of El Valle, *Gentry 6862* (MO); above El Potroso, *Sytsma 1766* (CAS, LL, MO); Valle de Antón, La Mesa, *Lewis et al. 2534* (COL, MO, UC); Cerro Gaital, north of El Valle, about 6 km N of the northern edge of the town, on the N-facing slope of Cerro Gaital, *Reveal & Balogh 4954* (US); Las Minas, *Sytsma 4033* (CAS, LL, MO); Valle de Antón, La Mesa, *Wilbur et al. 11149* (DS, F, G, GH, IJ, LL, MICH, MO, NY, TEX).—COLÓN: Canal Zone, Fort Clayton, vic. Río Piedras, along rd to Puerto Bello, 09°13′N,

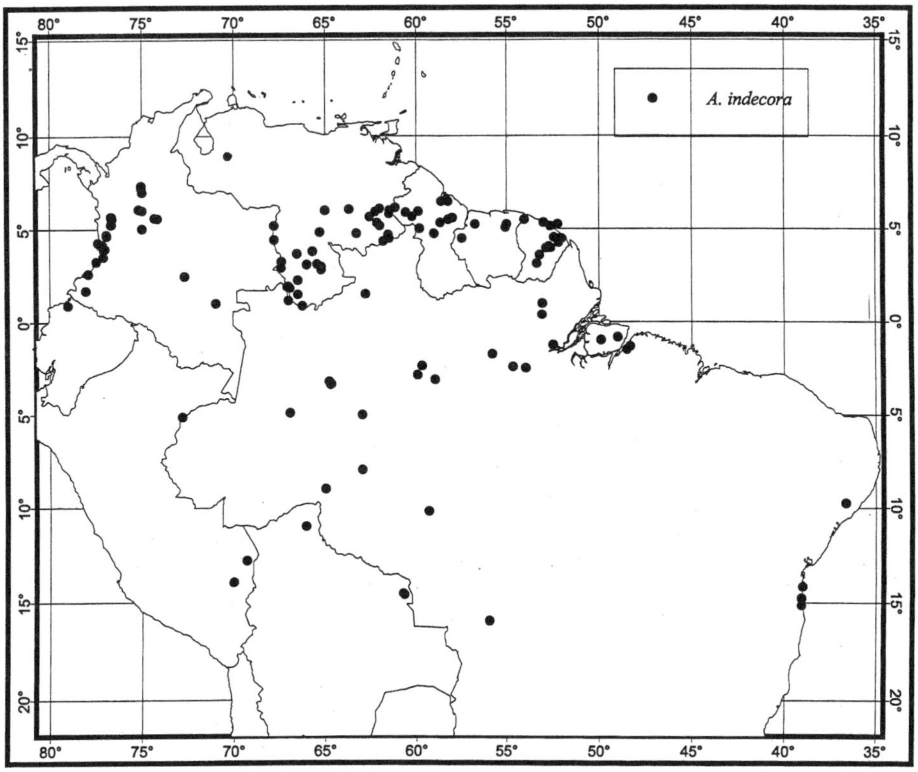

FIG. 33. Distribution of *Aciotis indecora* in South America.

79°45′W, *Blum et al. 2523* (MO); San Miguel de la Borda, *Croat 9893* (F, MO), *Croat 9993* (F, MO, NY, UC); Guasimo, *Croat 9956* (MO); Santa Rita Ridge, *Croat 13179* (MO); along road ca. 1 mi from Boyd–Roosevelt hwy, *Croat & Porter 15295* (MO); Río Indio–Miguel de la Borda rd, *Croat 36905* (MO); Río Guanche, *D'Arcy 10191* (MO); María Chiquita, E of Río Piedras toward Portobelo, *Dwyer 7803* (GH, MO, UC); Santa Rita Ridge 8 km E of main hwy, *Dwyer 8998* (F, MO, NY), Coclé del Norte, *Hammel 4496* (MO); Santa Rita Ridge rd, Km 22–25 from Transisthmica Hwy, 09°26′N, 79°37′W, *Knapp et al. 1687* (LL, MO); 10 mi SW of Puertobelo, *Liesner 1095* (AAU, COL, K, LL, MO, RB, WIS); 6 mi SW of Portobelo, *Luteyn 1412* (MO, US); Sabanita, *Medina 10* (S); Portobelo–Pilón, bridge over Río Piedras, *Nee & Mori 3644* (MO, WIS); Sabanita–Portobello, 5 mi NE, *Wilbur & Luteyn 11617* (F, GH, MO, US); wooded slopes about 5 mi from highway at Sabanita, *Wilbur & Luteyn 11840* (F, GH, LL, MICH, MO, NY).—DARIÉN: Cana, *Croat 37969* (MO); SW of Río Chico, ca. 10 km upstream from Nazareth, 08°15′N, 77°35′W, *Hahn 154* (MO); Cana, base of Cerro Pirre, *Hamilton & Stockwell 1425* (CAS, MO).—PANAMÁ: along road to Chagres Valley, *Almeda et al. 5866* (CAS); El Llano–Cartí Rd, *Busey 378* (F, MO); Cerro Jefe–La Eneida, *Croat et al. 13086* (F, MO, NY); Cerro Pilón, near Gamboa, *Croat 33719* (MO); Battelle Memorial Institute, Sea Level Canal Bioenviromental Program, *Duke 9407* (MO); Cerro Pilón, near El Valle, *Duke 12059* (MO); Cerro Azul, *Dwyer 1524* (MO); Goofy Lake–Cerro Jefe, *Dwyer 7088* (GH, MO); Cerro Campana, 3/4 of the way to the summit from the Pan-American Highway, *Dwyer et al. 4703* (MO); Cerro Azul–Cerro Jefe, *Elías & Hayden 1806* (MO, UC); Santa Rita, *Gómez-Pompa et al. 3307* (MEXU); Lago Cerro Azul, *Hamilton & Krager 3175* (MO); 5–6 mi from Pan-American Hwy on El LLano–Cartí road, *Knapp 2671* (CAS, MO); Cerro Campana–Capira, 8.6 mi SW, *Luteyn 962* (MO); altos de Campanas, *Méndez 33* (CAS, NY); above Su Lin Motel, *Porter et al. 4238* (MO, UC); Goofy Lake, *Sullivan 50* (CAS, MO); Campana–Motel Zulin, *Vergara & Torres 112* (MO); 8 km NE of Santa Rita, *Wilbur & Weaver 10842* (C, F, MICH, MO).—SAN BLAS: Nusagandi, El Llano–Cartí Road, *de Nevers & Núñez 3572* (CAS, MO); Cangandi, Río Cangandi, *de Nevers et al. 4921* (CAS, MO); Cangandi–Mandinga Rd, 2–5 mi S of Mandinga airport, *Duke 14794 (3)* (MO); El Llano–Cartí road, *McDonagh et al. 91* (BM, MO).—VERAGUAS: near Sta. María, road to Río

Calovébora above Escuela Agricultura Alto Piedra, *Antonio 2020* (MO); mouth of Río Concepción, *Lewis et al. 2813* (COL, MO, UC).

Colombia. AMAZONAS: Río Lagara–Parana (affl. Río Putumayo), coord. La Chorrera, territorio de indios Witoto Jitomagaro, familia Kuiru (17 km en aval de la Chorrera), *Gasche & Desplats 126* (US); Río Igara–Paraná, Bella Vista, *Sastre 2165* (COL).—ANTIOQUIA: Termales de Sto. Domingo, *Barkley & Bouthillette 30C539* (GH); Cocorná, 06°04′N, 75°11′N, *Bro. Daniel 1615* (US); Mpio. Gómez Plata–Yolombó, La Cancana, *Callejas et al. 2327* (HUA, NY, US); Anorí, *Denslow 2407* (COL, HUA, US, WIS); Zaragoza, *Denslow 2161* (WIS); Mpio. Anorí, Corregimiento Providencia, valle del R. Anorí, entre Dos Bocas y Anorí, *Fonnegra et al. 658* (HUA); Mpio. Sonsón, región Río Verde, Hac. La Soledad, a orillas de la Quebrada Curubital, *Gutiérrez 35550* (F, UC); Medellín–Pto. Triunfo, *Juncosa 991* (COL, MO, US); El Castrillón, San Luis–Antioquia, *Loaiza et al. 118* (HUA); Mpio. San Rafael, Morro Pan de Azúcar, *Orozco et al. 777* (COL); Vereda El Vizcocho, margen derecha del Río Bizcocho, *Orozco et al. 961* (COL); Botero, Cauca Valley, *Pennell 10977* (GH, NY, US); Agaria, *Sandeman 5571* (COL, K); Segovia, *Sandeman 5576* (COL, MA); vicinity of Medellín, San Roque, *Toro 1347* (NY); via a San Luis, Autopista Medellín–Bogotá, *Uribe et al. 44a* (MO); between Dos Bocas and Anorí, near Caño Tirana bridge, 07°21′N, 75°03′W, *Zarucchi 3278* (MO, NY, US).—CAUCA: La Seta, upper Cauca valley, *Lehmann 5488* (F, K); Valle de Cauca, *Triana 3881* (COL, E, NY); Cauca, *Triana s.n.* (BR, W).—CHOCÓ: La Oveja and Quibdó, *Archer 1712* (US); La Platina, Barrio Obrero, *Córdoba et al. 9* (MO); Puente de Cabí, *Córdoba et al. 45* (COL, MO); Quibdó–Istmina Rd, Km 6.6 S, 05°33′N, 76°37′W, *Croat & Cogollo 52151* (COL); Las Animas–Quibdó, Km 1, *Croat 55939* (MO); along road between Quibdó and Medellín, Km 9 W of Tutunendo, *Croat 56189* (MO); Pizarro, Pié de Pepé–Berrecul Rd, Km 21–24, *Espina & García 1922* (COL, MO); carretera Quibdó–Guayabal, 05°40′N, 76°38′W, *Forero et al. 1102* (COL, MO); Río Munguidó, afluente de Río Atrato, alrededores de Altagracia, *Forero et al. 1486* (COL, MO, NY); Quibdó–Tutunendo, Km 15, *Forero & Jaramillo 2532* (COL, MO); Hoya del R. San Juan, alrededores de Noanama, *Forero et al. 4566* (COL, MO, US), *Forero et al. 4567* (COL, MO, US); Palestina, *Forero et al. 3827* (MO); Quebrada La Sierpe, frente a Palestina, *Forero et al. 3885* (MO); Quebrada El Quicharo, *Forero et al. 4189* (MO, US); Docordó, *Forero et al. 4364* (MO, US); Río Fujiadó, *Forero et al. 4782* (MO, US); Itsmina, divorcio de aguas entre Hoyas de ríos Atrato y San Juan, 05°15′N, 76°38′W, *Forero & Jaramillo 5333* (COL, MO, US); Quibdó–Yuto Rd, Río Cabí, *García 34* (COL, MO); Quibdó–Guayabal, Río Hugón, *García et al. 79* (COL, MO); Villaconto, near Quito and Paimadó Rivers, *García-Barriga 11165* (COL, US); Medellín–Quidbó Rd, El Siete, *Gentry & Fallen 17896* (COL, MO); Quibdó, *Gentry & Renteria 23859* (COL, MO); Río San Pablo–Río Quito, Km 5 of Istmo de San Pablo, *Gentry & Renteria 24079* (COL, MO); La Platina, *Idrobo & Cuatrecasas 2706* (COL, F, NY, US); Frente a Barrio Obrero, *R. Palacios 4* (COL, MO); Río Quito, 5 km above junction with Río Atrato, *Prance 28034* (COL, NY, US); La Platina, *Uribe 3131* (NY).—CUNDINAMARCA: Caparrapí, Hda. Saldaña, *García-Barriga 7682* (COL, US); Yacopí, Insp. de Policía de Guadualito, vereda del Lamal, *Lozano et al. 7157* (COL).—EL VALLE: Buenaventura, *Pennell & Killip 5304* (GH, NY, US).—HUILA: E of Neiva, Cordillera Oriental, *Rusby & Pennell 500A* (NY).—NARIÑO: without locality, *Triana 6195* (BM [3 sheets], COL); Gorgona Island, E side, 02°36′N, 77°54′W, *Fernández 398* (COL, NY, US).—TOLIMA: Mariquita, Chorritos de San José, *Uribe 2850* (COL, F, NY).—VALLE DEL CAUCA: Bajo Calima Region, frente a Puerto Patiño, *Cabrera 6307* (MO); Bajo Calima Region, along road from Buenaventura to Málaga, *Croat 70089* (AAU, MO); Río Calima, La Trojita, *Cuatrecasas 16368* (F, NY); Río Cajambre, Barco, *Cuatrecasas 16954* (F, NY); Córdoba, *Cuatrecasas 19808* (F); Granja Agroforestal a orillas Río Calima, *Devia 440* (COL, MO, US); Restrepo, Chancos–Cusumbo, *Devia & Prado 2540* (CAS); Dagua River, *Killip 5069* (GH, NY, US); Santa Rosa, *Killip 11520* (GH, NY, US); about 18 km E of Buenaventura, *Killip & García 33257* (BM, COL, NY, US); Buenaventura, 10–15 km E of Buenaventura, *Killip 34922* (COL, NY, US); Puerto Merizalde, Río Nanay, *Gentry & Juncosa 40582* (COL, MO, US); Timba, *von Sneidern 1198* (S); Córdoba, *von Sneidern 5340* (NY, S, US).—VAUPÉS: Río Cuduyarí, affluent of Río Vaupés, *Cuatrecasas 6844* (NY); Casa Alvarez, Bocas Carurú, *Cuatrecasas 7010* (NY); Bajo Vaupés, Mitú, Cerro de Urania, *Gutiérrez & Schultes 976* (GH), Mpio. San José del Guaviare, en inmediaciones de Ciudad de Piedra, Serranía La Lindosa, carretera San José–El Capricho, *López et al. 822* (MO); Río Kananarí, Cachivera Palito, *Schultes & Cabrera 13164* (GH, NY, US); Mitú, 01°15′N, 70°13′W, *Uribe 2928* (COL, NY). **Venezuela.** AMAZONAS: Cerro Impacto, Campamento El Cafetal, *Agostini 1567* (MO); orillas del Río Orinoco, alrededores de San Fernando, *Araque & Barkley 18 V.169* (NY, US); Depto. Atabapo, Caño Iguapo, 15 km SE de La Esmeralda, *Aymard & Delgado 8074* (US); Río Atacavi (confluente del Río Orinoco), *Bunting et al. 3654* (US); Territorio Federal Amazonas, San Carlos de Río Negro, ca. 20 km S of confluence of R. Negro and Brazo Casiquiare, 01°56′N, 67°03′W, *Clark 7013* (GB, MO, NY, RB); San Carlos de Río Negro, *Delascio et al. 9324* (US); Atabapo, O de cerro Cucurital y E de cerro Yapacana, *Guanchez 1254* (US); Atabapo, Río Ocamo, alrededores de Misión Silesiana Sta. María de los Guaicas, *Guanchez 1952* (MO); ca. 8 km NE of San Carlos de Río Negro, 01°53′N, 67°01′W, *Liesner 8841* (US); Depto. Río Negro, Caño Baria, *Liesner 17032* (CAS, F, MO, NY, US); Alto Río Orinoco, along Río Padamo, 10 km above mouth, *Maguire & Wurdack 34741* (F, MO, NY, US, VEN); Ríos Pacimoni–Yatua, Casiquiare, Río Yatua, *Maguire et al. 41951* (NY, US); carretera Solano–San Carlos, Km 6–7, *Morillo 5487* (VEN); Depto. Atures, Sta. Rosa de Ucata, *Romero & Guanchez 1800* (CAS, COL, GH, TEX); Depto. Río Negro, San Carlos del Río Negro–Solano, 01°50′N, 66°55′W, *Stergios & Aymard 4413* (CAS); Medio Casiquiare, Piedra La Esterita, desde la piedra arriba del Raudal Cabrua hasta la boca del Pasiba, *Stergios & Aymard 7646* (MO); orillas del Medio Casiquiare, entre boca del Caño Atamoni y 1 km abajo de la piedra Esterita, *Stergios et al. 8298* (AAU, MO); Bajo Emoni, entre la Laguna y tres vueltas mas arriba, *Stergios & Aymard 9272* (MO, NY); Río Siapa, *Stergios & Aymard 9235* (AAU); Yavita, Pimichin trail near Yavita, *Wurdack & Adderley 42904* (GH, NY, US).—BOLÍVAR: Campamento Indígena Haymary, *Aymard et al. 344* (MO); región de los ríos Icabaru, Hacha y cordillera sin nombre, a 280° de las cabeceras del Río Hacha, *Bernardi 2911* (NY); Icarabú–Santa Elena rd, Km 26–52 NE, *Croat 54207* (MO); 90 km de El Dorado hacia Sta. Elena, *Foldats 2729* (NY); Salto Danta, 125–126 kms al S de El Dorado hacia Sta. Elena, *Holst et al. 2110* (US); upstream of Río Pulpul, above Salto Pulpul, on plateau S of the peak of Uaipan-tepui, *Koyama & Agostini 7533* (NY); Dtto. Piar, Amaruay-tepui, 0.1–0.3 km N of SW corner of tepui, *Liesner & Holst 20510* (MO, TEX, US); orillas de la carretera entre Icabarú y Quebrada Negra, antes de El Paují, Gran Sabana, Dtto. Piar, 04°30′N, 61°30′W, *Ruiz-Terán & López-Palacios 10957* (F, US); Río Tonoro, quebrada del Trueno, *Stergios 10415* (US); Chimantá Massif, Amurí-tepui, *Steyermark & Wurdack 1331* (F, NY); Salto de Pacairao, ca. 3 km E to NE of Sta. Teresita de Kavanayén, *Steyermark 60524* (F, GH, NY, US); Cabeceras de Río Chicanang, 80 km (en línea recta) al suroeste de El Dorado, *Steyermark 89592* (F, NY, US); Sierra Ichún, Salto María Espuma (Salto Ichún), tributario del Río Paragua, *Steyermark 90422* (US); afluente E del Río Venamo, *Steyermark et al. 92650* (NY, US, VEN); Río Bonita (afluente de Río Cicuta y Río Nichare), en cerro hacia Sierra Maigaulida y Sierra Cervatana, *Steyermark & Gibson 95774* (US); Quebrada Los Brasileros–camino hacia Playa Blanca y Río Uaiparú, al SO de Icabarú, *Steyermark et al. 117721* (US); El Dorado–Sta. Elena road, 04°40′N, 61°33′W, *Wurdack et al. 2847* (NY, US, VEN).—TRUJILLO: Boquerón–La Gira, Dtto. Betijoque, 09°18′N, 70°40′W, *Pérez 63* (F). **Guyana.** Moraballi, Essequibo river, *Atkinson 75* (BM); Ituni, 35 mi S of Mackenzie, *Cowan 39226* (NY, US); Malali, Demerara River, 05°35′N, 58°00′W, *de la Cruz 2691* (MO); Kamakusa, Upper Mazaruni River, *de la Cruz 2800* (MO); Pomeroon River, *de la Cruz 3178* (MICH, MO, US); Kaieteur Falls, Potaro River, *de la Cruz 4455* (MO); Maipuri Falls, Karowrieng River, 05°41′N, 60°14′W, *Gillespie 2893* (BM, HB); Ransaroop Farm resort, 2 km E of Timehri Airport, 06°30′N, 58°15′W, *Hoffman & Capellaro 859* (HB); Arawak Amerindian land, 2 km up Pokerero River, above junction with Warakabra River, *Hoffman & Capellaro 925* (HB, INPA); Siparuni River; Pakutau Falls, 04°45′N, 59°01′W, *Hoffman et al. 1457* (HB); Ituni–Kwarkwani rd, Km 27, 05°30′N, 58°14′W, *Gillespie et al. 3013* (CAS, US); Moraballi Creek, near Bartica, *Sandwith 56* (K); Kamarang River, base of Utschi Falls, 05°54′N, 60°35′W, *Tillett & Tillett 45742* (G, NY, US); without locality, *Tutin 514* (BM); Kaletuk Gorge, *Tutin 534* (BM, US). **Suriname.** Jodensavanne–Mapane kreek, Suriname River, Blakawatra, *Boerboom 9172* (COL, US); Maratakka River, 20 km S of Saparra Creek, *Bosbeheer 10861* (US); Zanderij, Troelie, *Kramer & Hekking 2233* (NY); Tafelberg, Kappel-savanna, *Kramer & Hekking 3257* (NY); Moengo tapoe, *Lanjouw & Lindeman 978* (K, NY); Kabalebo Dam Project, distr. Nickerie, *Lindeman et al. 399* (A, AAU, COL, F, K, MO, NY, RB, S, US, Z); Zanderij I, *Maguire & Stahel 25046* (F, GH, NY, US); forest of Zandery, *Samuels 268* (A, GH, K, NY); Para District, *Wullschlägel 786* (BR); Karthey Lagoon, *Wullschlägel 1759* (BR). **French Guiana.** Montagne Cacao, S. de Cayenne, *Cremers 6928* (CAY, US); Montagne de Kaw, *Cremers 6948* (CAY, US); Riv. Mana, Saut Dalles, *Cremers 7252* (CAY); Station des Nouragues, Arataye, *Cremers 10893* (CAY, US); Savane Roche du Quatorze Juillet–Bassin du Bas–Oyapock, *Cremers 12305* (CAY, US); Crique Jean-Pierre, *Cremers et al. 13533* (MO); along St. Elie track, (W from Km 105 N of Sinnamary), NW of Sinnamary, *Croat 53812* (CAY); Arataye, rives de la rivière Arataye (affluent de l'Approuague) au saut Pararé, *de Granville 15* (CAY, US); Saül, sommet Sud du Pic Matecho, 30 km NE de Saül, *de Granville 3367* (CAY, US); route de Bélizon, Km 7, Eau Claire, *de Granville 4462* (CAY, US); piste de St. Laurent a Paul Isnard, PK-75, *de Granville 5346* (BR, CAY, US); route de Paul Isnard a St. Laurent, Km 40, *de Granville 5353* (BR, CAY, INPA, US); montagnes de la Trinité, *de Granville et al. 6163* (CAY, NY, US); 1.5 km env. au NE du Mont Belvedere, *de Granville 7118* (CAY, US); versant N de la montagne de Kaw, PK-46, *de Granville 9091* (CAY, US); station des Nouragues–Bassin de l'Approuague–Arataye, abords de la Crique Nouragues, *de Granville et al. 10358* (CAY); Montagne Cacao, *de Granville et al. 12705* (BR, CAY, MO); Rivière Petite Ouaqui, Saut Baille-Nom, *de Granville B.4976* (CAY); Crique Gregoire, station hydrologique ORSTOM, *Deward 11* (CAY); piste de Saint Elie, *Larpin 282* (CAY); Grégoire, Station no. 2, *Lescure 1* (CAY, US); near Kourou, *Renner 2144* (AAU); Saut Cascade, rivière des Cascades, *Sastre & Sastre 339* (CAY, US); St. Laurent, *Schnell 11380* (US); piste de Saint Elie, Arbocel, Km 16-2, *Toriola 1046* (CAY); Godebert, *Wachenheim 33* (BM, K); without locality, *Wachenheim 104* (BM). **Ecuador.** ESMERALDAS: San Lorenzo, Parroquia Ricaurte, Centro Pambilar, 01°08′N, 78°36′W, *Aulestia 876* (MO). **Peru.** CUSCO: 17 km W of Pilcopata, ca. 13°03′S, 71°10′W, *Gentry et al. 23662* (F, MO, NY, US); Paucartambo, Asunción, 14°29′S, 71°27′W, *Marín 1702* (F).—

LORETO: Upper Río Nanay, *Williams 1180* (F, NY).—MADRE DE DIOS: banks of Río La Torre, along the Río Aguas Negras, part from 4–15 km up river from Explorer's Inn, near confluence of Río Tambopata and Río La Torre, 39 km SE of Puerto Maldonado, 12°50′N, 69°20′W, *Smith et al. 1633* (NY, US). **Bolivia.** EL BENI: Vaca Diez, Riberalta, 11°00′S, 66°05′W, *Solomon 7592* (MO, NY, US).—SANTA CRUZ: Velasco, Parque Nacional Noel Kempff, El Encanto, 14°37′S, 60°42′W, *Saldías et al. 2702* (MO). **Brazil.** ALAGOAS: Arapiraca, 09°45′N, 36°39′W, *de Lyra et al. 155* (UEC).—AMAPÁ: Mapati, Mpio. Macapa, *Black & Lobato 50-9699* (NY); Serra do Navio, Río Amapari,01°N, 53°05′W, *Cowan & Maguire 38057* (GH, NY, RB); Río Jari, Cachoeiras da Guaribas, 00°24′N, 53°07′W, *Egler & Irwin 46452a* (F, G, K, MG, NY, US); S. Joaquim do Pacui–Macapa, *Rabelo 565* (MG, NY); Río Pedreira, medio curso–Macapa, *Rabelo et al. 2655* (NY, US); Matapí, *Ribeiro 1611* (NY, RB, US).—AMAZONAS: Lago de Castanho-Mirim, estrada da Petrobrás, *Albuquerque et al. 709* (INPA); Mpio. Tefé, Río Solimões, lago Tefé, *Amaral et al. 724* (INPA, K, MG, NY, US); estrada Manaus–Porto Velho, *Bahia 88* (US); Coari, *Bahia 148* (F, NY, US); Antonio Lemos, varzea Igarapé, Inst. Agron. do Norte, Río Negro, *Black 48-2977* (RB); Ilha de Marajó, caminho de Jubim para Condeixa, *Black 48-3552* (RB); Dtto. Agropecuario, Reserva 1501 (Km 41) of the WWF/INPA MCS Project, 02°24′S, 59°43′W, *Boom et al. 8709* (AAU, NY); Manaus–Porto Velho highway, Km 215, *Campbell et al. P20793* (F, INPA, K, MO, NY, S, U, US); São João, farm of Sr. Taiyo, Maués, *Campbell et al. P22085* (F, INPA, NY, U, US); Mpio. Tefé, atrás da cidade de Tefé, *Cid Ferreira & Lima 3253* (INPA, K, NY, US); Mpio. Alarães, Río Solimões, estrada de Alarães para a Vila Nogueira, 03°14′S, 64°48′W, *Cid Ferreira & Lima 3822* (F, GH, INPA, K, MO, NY, US); Manaus, campina das Pedras, BR 174, Km 115, próximo à margem do Rio Tarumãzinho, *Corrêa et al. 101* (INPA); Carauari, Poço Juruá, *da Silva et al. 592* (MG, US); Mpio. Humaitá on plateau between R. Livramento and R. Ipixuna, basin of Rio Madeira, *Krukoff et al. 7292* (BM, BR, F, G, GH, K, M, MICH, MO, NY, RB, S, US); Rio Javari, behind Palmeiras Army Post, *Lleras et al. P17019* (INPA, K, NY, US); Km 405 on Manaus–Humaitá road, *Lowrie et al. 46* (INPA, NY); Rio Negro, I. Isabel, *Luetzelburg 22110* (M); Tunuy, R. Içana, *Luetzelburg 22754* (R); Rio Negro, Ilha Acaburu, *Maia et al. 420* (INPA); Lago de Jaradá, Rio Cuieiras, ca. 40 km from mouth at the Rio Negro, *Mori et al. 20362* (INPA, NY, US); Manaus, Cachoeira do Tarumã, *Pereira 3485a* (RB); Serra da Lua, Rio Urubu, between Cachoeira Iracema and Natal, *Prance et al. 5000* (K); vicinity of Pico Rondón, Perimetral Norte highway Km 211; Rio Curuquetê, vicinity of Cachoeira Santo Antonio, *Prance et al. 14346* (INPA, NY, US); 3 km from Km 211, *Prance et al. 28747* (INPA, NY); Manaus, igarapé da Cachoeira Alta do Tarumã, *Rodrigues & Chagas 2296* (INPA, US); Manaus, estrada da Forquilha, *Rodrigues & Chagas 4564* (INPA, US); estrada Manaus–Porto Velho, trecho Castanho–Tupana, *Silva et al. 154* (INPA); Reserva Forestal Ducke, Manaus–Itacoara, Km 26, Igarapé do Tinga, *Souza 360* (INPA); Reserva Forestal Ducke, Floresta de Baixio, *Souza 1089* (INPA).—BAHIA: Partie Meridionale, *Blanchet 1030* (G); Mpio. Una, Reserva Biol. do Mico–Leão (IBAMA), entrada no Km 46 da Rod. BA-001 Ilhéus/Una, picada paralela do Rio Maruim, *Amorim et al. 1375* (CAS, MBM, MO, NY); Maraú, *Belém 1808* (US); margem de Igarapé, *Belém 3482* (NY, US); Mpio. Ilhéus, entrada pelo Banco da Vitória, ca. 3.5 km NE da sede, *Carvalho et al. 4602* (UEC); Mpio. Una, Reserva Biol. Mico-leão (IBDF), Km 8 da rodovia Una/Ilhéus (BA–001), *dos Santos & Alves 137* (US); Ba. Uruçuca, Saida para Taboquinha, *dos Santos 642* (RB, US); Ilhéus, *Luschnath s.n.* (BR); Mpio. Una, Estação Experimental Lemos Maia, CEPLAC, *Rylands & Hage 178* (INPA, US); Mpio. Ilhéus, Cidade de Ilhéus, 3 km N of Rodoviaria, Mata da Esperança, 14°47′S, 39°04′W, *Thomas et al. 10513* (NY, UEC, US); Ilhéos Repartimento (Rib. da Fortuna), *Velloso 798* (R); Ilhéus, Pirataquissé, *Velloso 1039* (GH, R).—MATO GROSSO: Humboldt Center to port on Rio Aripuanã, 10°12′S, 59°21′W, *Berg et al. P18575* (AAU, INPA, K, MG, MO, NY, R, S, US); Estação Ecológica de Iquê-Juruena, *Guarim Neto & Neto 528* (INPA); Sta. Cruz, *Moore 471* (BM, E, NY, R).—PARÁ: Estação Ecológica do Jari, Projeto Reserva Genética SEMA, *Beck et al. 128* (INPA, NY, US); Rio Trombetas, 3 km S of Cachoeira Porteira, *Campbell et al. P22572* (INPA, NY, US); Belém, Horto Botânico (Prov. R. Purús, Bom Lugar), *Huber MG-4975* (HB, US); Bragança, Colonia Benjamin Constante, estrada Sul, Km 8 (Propiedade de Sr. M. L. Ferreira), *Jangoux & da Silva 1208* (MG, US); near Antonio Lemos, *Krukoff et al. 5895a* (GH, K, NY); Tome Aqu, perto fazenda Borba Gato, *Nascimento 369* (MG, NY, US); Belém, João Coelho, Ig. Caraparú, *Pereira & Egler 3341a* (HB); Belém, IPEAN, Reserva APEG, *Pires & Silva 10482* (US); Mpio. Benevides, Fazenda Maratá, about 30 km ENE of Belém, on road to Benfica, *Plowman et al. 8062* (F, INPA, MG, MO, NY, US); Ilha de Marajó, Rio Anajás, logo acima de Anajás, 00°59′S, 49°55′W, *Prance et al. 30191* (F, GH, INPA, MO, NY, US); Ilha de Marajó, Rio Anajás, *Rabelo et al. 3677* (CAS, INPA, MO, NY); route de Belém–Mosqueir, *Sastre 140* (US); Rio Solimões, Sto. Antonio do Ica, estrada p/o Cemitério, *Silva 2123* (MG); Santarém, *Spruce 977* (G); Santarém, estrada que liga Alter do Chão, *Vilhena et al. 264* (NY, US).—RONDÔNIA: Velho–Guajará-Mirim, Sub-base Proj. RADAM–Estrada do Palheta–Fronteira Brasil/Bolivia, *Cordeiro 842* (US); on road to Saldanha, close to Guajará-Mirim, *Kirkbride et al. 2703* (INPA, K, NY, US); trail north of Rio Madeira from 2 km below confluence of Rio Abunã, *Prance et al. 8326* (INPA, NY, US); Vilhena, Rio Piracolina, *Silva & Rosario 4922* (MG, NY).—WITHOUT LOCALITY: *Riedel s.n.* (K, M, W); *Riedel s.n.* (BR, C, F, G, M, MO, NY, S, US, W, Z); *Wawra & Maly s.n.* (W).

Aciotis indecora is characterized by the dense glandular pubescence on the entire plant and the baccate, caducous fruits. In most collections the abaxial surface of the petals and the apex of the ovary are beset with glandular trichomes.

Most Mexican collections have inflorescences with three to four pairs of first order paraclades, and those from the Central and South America have up to five pairs. Also, most of the Central American collections have glabrous ovaries; yet, *Molina 14949* from Nicaragua has the ovary apex glandular-pubescent. As a rule the elongated anthers are approximately the same size in both whorls; however, *Silva et al. 592* (Brazil) has the anthers of the inner whorl half as long (0.4 mm) as those of the outer whorl (0.8 mm). In the majority of specimens pedicels as well as the fruits are caducous; *Dwyer 7803* (GH, UC) and *Lewis 2813* (UC) from Panama are exceptional in having old inflorescences with persistent pedicels.

Aciotis indecora may hybridize with *A. purpurascens* and *A. rubricaulis*; see Appendix 2.

Mexican and Central American collections with glabrous petals were previously recognized as *A. rostellata*, but *Zamudio & Magaña 92* and *Orcutt 6514* have the petals abaxially glandular-pubescent. The type of *Spennera acutiflora* from Brazil has glabrous petals but also a paniculate inflorescence and the ovary apex glandular-pubescent.

Specimens that have the stems and petioles densely covered with eglandular trichomes and the ovary apex usually glandular-pubescent have been segregated as *A. laxa* (e.g., *Kubitzki & Poppendieck 79-22*). These variants seem similar to *A. rubricaulis*, but they can easily be distinguished by their (usually) cordate leaves, glandular-pubescent ovary apex, and abaxially glandular-pubescent petals.

Several illustrations of *A. indecora* show some of the stable and variable characters. Cogniaux's plates (1885, pl. 107 and 108, as *A. acutiflora* and *A. laxa* var. *robusta*) show cuneate leaves and old inflorescences that have shed the fruits, features typical for *A. indecora* but not *A. viscida*, a somewhat similar species. Bonpland's color plate (1821, pl. 49, as *Rhexia indecora*) shows cordate leaves and glandular-pubescent stems, petioles, and hypanthia. The dimorphic stamens observed in some other Brazilian individuals are also shown. Brade (1958, as *A. anomala*) illustrated the paniculate inflorescence (the fruits already dispersed) and the glandular-pubescent ovary apex.

In the protologue for *A. indecora* var. *macrophylla*, Cogniaux (1885) listed as syntypes *Poeppig s.n.* (W!) and *Lechler 2404* (G! K!). *Poeppig s.n.* is here designated the lectotype; *Lechler 2404* is *A. rubricaulis.*

Whiffin and Tomb (1972), for their study of seeds of Neotropical capsular-fruited species of Melastomataceae, included *A. indecora* as represented by *Contreras 6545* (LL, Guatemala) and *Prance et al. 8326* (US, Brazil).

Concoctions prepared by boiling branches and leaves of *A. indecora* are drunk in Belize in order to prevent nausea in early pregnancies (*Balick et al. 2607*). *Yuyito* is used for making herbal tea (*Zarucchi 3278*).

13. Aciotis viscida (Bentham) Freire-Fierro in Berry et al., Fl. Venez. Guayana 6: 276. 2001. *Spennera viscida* Bentham, J. Bot. (Hooker) 2: 296. 1840. *Aciotis indecora* var. *glabrescens* Cogniaux in Eichler, Fl. bras. 14(3): 470. 1885.—TYPE: GUYANA. Without locality, 1839, *R. H. Schomburgk I 999* (holotype: K; isotypes: BM! BR! E! G [2 sheets]! K! NY [2 sheets]! US! W!).

Herbs 0.4–1 m tall; stolons absent; young stems slightly quadrangular with inconspicuous wings, glandular-pubescent, the trichomes <1 mm long. Leaf blades (2–) 3–5.5

(–6.7) cm long, 1.5–2.5 (–3.2) cm wide, ovate, apex acute, base obtuse to cordate, margin serrulate, papyraceous, 5–7-veined; adaxially dark green, pubescent with glandular and eglandular trichomes, abaxially green, pubescent with glandular and eglandular trichomes, without web-like striations; petiole 0.5–1.4 (–3) cm long, with narrow wings, glandular-pubescent, the trichomes <1 mm long. Inflorescence a thyrse, 5.5–9 cm long, 2–4 (–5) cm wide, with (3–) 4–6 pairs of first order paraclades; rachis glandular-pubescent; internodes 1.5–2.2 cm long; first order hypopodia (0.3–) 0.8–1 (–1.2) cm long, biparous; second order hypopodia 0.3–0.6 cm long, biparous; third order hypopodia 0.2–0.4 (–0.6) cm long, uniparous; fourth order hypopodia ca. 0.2, uniparous; terminal paraclade 1–2.2 cm long, with 2 flowers per cm; bracts ca. 0.8 mm long, ca. 0.2 mm wide, ovate, apex acute, abaxially glandular-pubescent. Hypanthium ca. 2 mm long, ca. 1.5 mm in diameter, terete, without filiform sclereids in parenchyma, abaxially densely glandular-pubescent, adaxially glabrous. Petals 2–2.8 mm long, 0.9–1.2 mm wide, ovate, apex acute, base truncate, with a terminal glandular trichome, abaxially with glandular trichomes. Stamens with filaments 1–2.5 mm long, anthers 0.8–1.2 mm long, pedoconnective 0.2–0.5 mm long. Ovary without filiform sclereids in parenchyma, apex glandular-pubescent. Fruit a capsule, pericarp and placenta persistent after dehiscence, ca. 2.8 mm long, ca. 2 mm in diameter, bilocular, pericarp coriaceous, not translucent; seeds ca. 0.50 mm long, ca. 0.40 mm wide. Fig. 34.

Phenology. Flowering and fruiting throughout the year, except in February.

Distribution (Fig. 18). Amazon Basin; flooded or non-flooded open areas; 50–250 m.

ADDITIONAL SPECIMENS EXAMINED. **Colombia**. VAUPÉS: Río Apaporis, entre los ríos Cananari y Pacoa, 00°07′N, 71°15′W, *García-Barriga 13850* (COL, NY, US). **Venezuela.** AMAZONAS: San Carlos de Río Negro, *Liesner 3596* (MO, US); Depto. Casiquiare, Yacami, 35 km ENE of San Carlos de Río Negro, 01°58′N, 66°45′W, *Liesner & Delascio 21870A* (MO, NY, US); Río Cuao, Río Orinoco, *Maguire & Politi 27366* (F, K, NY, RB, US); Piedra Guachapita y laguna de Catárapo, 02°S, 66°55′W, *Stergios & Aymard 7392* (MO, US); Dtto. Alto Orinoco, La Esmeralda, 200 m SW of La Esmeralda, *Zimmermann 37* (US). **Ecuador**. NAPO: R. Lagartococha, near Redondo Cocha and Imuya Cocha, 2 hours in canoe from Río Aguarico, 00°35′S, 75°15′W, *Laweson et al. 44243* (AAU, MO). **Peru.** LORETO: Quebrada de Nauta, along Río Marañón, 04°29′S, 73°35′W, *Croat 17510* (F, GH, MO, XAL); Quebrada Nawampa, near mouth of Río Tigre, *Croat 17567* (MO, US); San Regis, cerca de confluencia de ríos Marañón y Ucayali, *Ferreyra 5133* (US). **Brazil.** AMAZONAS: Lago de Castanho-Mirim, *Albuquerque et al. 709* (INPA); Rio Negro, between Manaus and São Gabriel, Ilha Acaburu, 00°25′S, 66°25′W, *Alencar 420* (NY, US); Rio Urubu, Iracema waterfall, *Calderón et al. 2962* (INPA, K, MO, NY); Dtto. Agropecuario-Reserva 1501, Km 41 do Smithsonian/INPA, *Freitas 159* (NY); Mpio. Humaitá, near Tres Casas, *Krukoff et al. 6456* (BM, BR, F, G, GH, MICH, MO, NY, S, US); I. Isabel Novo, *Luetzelburg 21110* (R); Ilha Acaburu, *Maia et al. 420* (INPA); Río Negro, Maribi, *Martius s.n.* (M); Ega, *Poeppig 2496* (G, NY, W); Manaus–Itacoatiara rd, Km 212, Rio Urubu, 03°08′S, 59°00′W, *Prance et al. 4899* (COL, F, K, NY, R, S, US); Rio Urubu, Manaus-Caracarai, Km 115, *Prance et al. 23586* (MG, NY, S, US); Manaus-Caracarai rd, Km 64, ZF-3 rd, Km 34, *Renner 919* (AAU, INPA, US); São Gabriel, ca. 00°07′S, 67°05′W, *Spruce 2331* (K); Sta. Isabel, *Tate 93* (NY).—BAHIA: entre Bom Gosto e Olivença, 14°57′S, 39°01′W, *Fróes de Lemos 20031* (NY).—CEARÁ: without locality, *Gardner 999* (G).

Aciotis viscida is very similar to *A. indecora* but differs having narrower inflorescences and capsular fruits with a coriaceous pericarp and persistent placenta. In its foliage, pubescence, and capsular fruits it also resembles *A. polystachya*; however, that species is smaller (less than 50 cm tall), and has double biparous cymes and filiform sclereids.

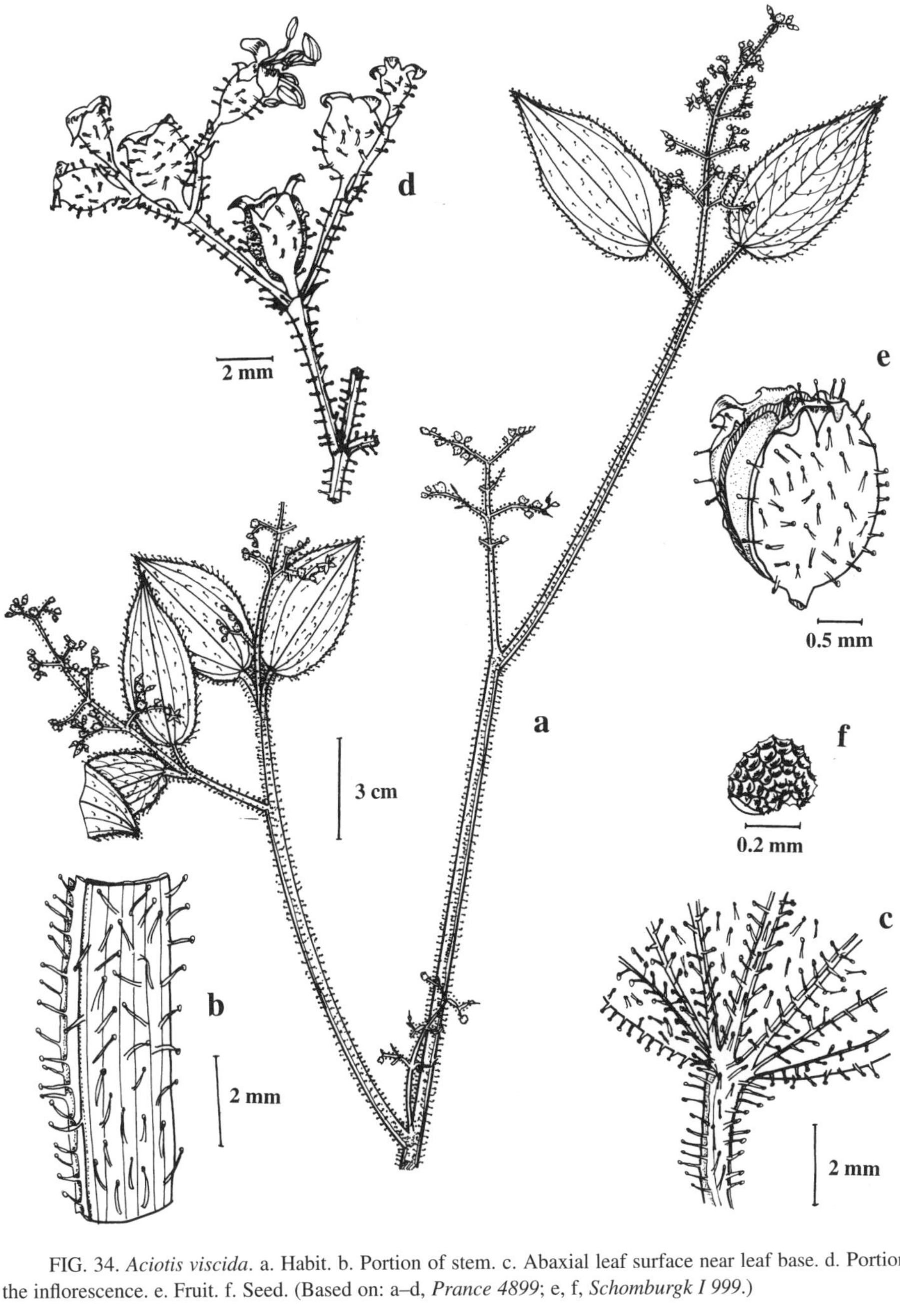

FIG. 34. *Aciotis viscida*. a. Habit. b. Portion of stem. c. Abaxial leaf surface near leaf base. d. Portion of the inflorescence. e. Fruit. f. Seed. (Based on: a–d, *Prance 4899*; e, f, *Schomburgk I 999*.)

Doubtful and Excluded Names

Clidemia herbacea Steudel ex Cogniaux in Eichler, Fl. bras. 14(3): 468. 1885, pro syn.

Melastoma annuum Schrank ex DC., Prodr. 3: 115. 1828, pro syn.

Melastoma aquatica Aublet, Hist. Pl. Guiane 1: 430, pl.169. 1775. *Rhexia aquatica* (Aublet) Swartz, Fl. Ind. Occid. 2: 650. 1797. *Spennera aquatica* (Aublet) DC., Prod. 3: 116. 1828. *Aciotis aquatica* (Aublet) D. Don ex Loudon, Loud. Hort. Brit. 174. 1830.—TYPE: FRENCH GUIANA. Between Cayenne and Courou, *Aublet s.n.* (holotype: BM). = *Nepsera aquatica* (Aublet) Naudin.

Melastoma cordata Vellozo, Fl. flum. 178. 1829 ["1825"], t. 4, 114. 1831 ["1827"]. *Aciotis cordata* (Vellozo) J. F. Macbride, Publ. Field Columbian Mus., Bot. Ser. 4: 175. 1929 (the combination also proposed a second time by Macbride; Candollea 8: 23. 1940).—TYPE: BRAZIL. Rio de Janeiro: *Vellozo s.n.* (holotype: not located).—The protologue does not provide enough information to fix the name. The accompanying plate shows a plant with a terminal inflorescence and a dissected flower with ten stamens, five sepals, and six petals. *Aciotis* has only eight stamens, four sepals, and four petals; the plate of *Melastoma cordata* does not represent any species of *Aciotis*.

Melastoma fragile ["*fragilis*"] Richard ex DC., Prodr. 3: 115. 1828, pro syn.

Melastoma laxa Richard ex DC., Prodr. 3: 116. 1828, pro syn.

Melastoma pellucens Richard ex DC., Prodr. 3: 115. 1828, pro syn.

Melastoma rubricaule Schrank ex DC., Prodr. 3: 116. 1828, pro syn.

Melastoma uliginosa Bonpland ex Naudin, Ann. Sci. Nat. Bot., sér. 3, 14: 144. 1850, pro syn.

Melastoma viscosa Bonpland ex Naudin, Ann. Sci. Nat. Bot., sér. 3, 14: 146. 1850, pro syn.

Rhexia hexandra Schrank ex DC., Prodr. 3: 115. 1828, pro syn.

Rhexia pendulifolia Bonpland, Descr. Pl. Malmaison 68, pl. 26. 1815. *Spennera pendulifolia* (Bonpland) DC., Prodr. 3: 116. 1828. *Nepsera pendulifolia* (Bonpland) Naudin, Ann. Sci. Nat. Bot., sér. 3, 13: 29. 1850. *Aciotis pendulifolia* (Bonpl.) Triana, Trans. Linn. Soc. London 28: 52. 1871.—TYPE: GUYANA. Without locality, *Bonpland s.n.* (holotype: P).—According to the protologue and original plate, the type has stems with short trichomes, serrate-crenulate leaves, and small, white, and pentamerous flowers. Candolle (1828) states that this species has five sepals, ten stamens, and three locules; therefore it could belong to another genus of Melastomataceae but not *Aciotis*.

Spennera lancifolia Steudel ex Cogniaux in Eichler, Fl. bras. 14(3): 475. 1885, pro syn.

Spennera sieberi Steudel, Flora 27: 720. 1844. *Aciotis sieberi* (Steudel) Triana, Trans. Linn. Soc. London 28: 52. 1871.—TYPE: TRINIDAD. Without locality, 1822, *Sieber 156* [*Wrbna s.n.*] (holotype: P) = *Nepsera aquatica* (Aublet) Naudin, fide Wurdack et al. (1993).

ACKNOWLEDGMENTS

I thank the curators of the following herbaria, who kindly sent material of *Aciotis* on loan to MO (acronyms according to Holmgren et al., 1981): A, AAU, AMES, BHCB, BM, BR, C, CAS, CAY, COL, CR, DS, E, F, G, GB, GH, GUA, HB, HRB, HUA, IJ, INPA, K, LL, LPB, M, MA, MBM, MEXU, MG, MICH, MO, NY, QCA, QCNE, R, RB, S, SPF, TEX, TRIN, UC, UEC, US, VEN, W, WIS, XAL, and Z. I am grateful to Susanne Renner, Robert Ricklefs, Phil Fraundorf, Haresh Siriwardane from the University of Missouri-St. Louis, as well as Richard Keating, Gerrit Davidse, and Gordon McPherson from the Missouri Botanical Garden, for their instruction and help; to Frank and Mary Beth Almeda for their generous hospitality during my stay in San Francisco; to Frank Almeda, Fei-Mei Chuang, and the staff at the California Academy of Sciences for their patience and help during my visit; to Jim Zarucchi, Charlotte Taylor, and Roy Gereau for their comments on the manuscript; to Jim Solomon and the staff of the Missouri Botanical Garden for help with the loans and all other logistic matters. I also thank Dra. Laura Arcos Terán of the Pontificia Universidad Católica del Ecuador for her generosity, and my colleagues and the staff at Herbario Nacional del Ecuador for their friendship. José Fernando Baumgratz kindly sent me a copy of one of Vellozo's plates, which is deposited at the Jardim Botânico do Rio de Janeiro, Brazil. Two anonymous reviewers offered thoughtful suggestions, and Christiane Anderson provided invaluable help with manuscript preparation.

I am deeply grateful to my wonderful husband, Joseph C. Welling for his moral and economic succor, especially during 1998. I am indebted to the Welling family, especially Jean Bollinger and Gail Bowman, the Freire-Fierro family for their support and understanding, and the López Family for the assistance in the field in Ecuador. Also, I thank all my fellow graduate students at the Missouri Botanical Garden, especially Kathy Aldás, Jason Bradford, Tibisay Escalona, Anthony Jardim, Susana León, Rosa Ortiz-Gentry, Thomas Prinzie, and Armand Randrianasolo for their altruism and friendship.

The work for this monograph was partly funded by a two-year fellowship provided by FUNDACYT (Fundación para la Ciencia y la Tecnología del Ecuador)-LASPAU, and carried out at the Missouri Botanical Garden and the University of Missouri-St. Louis. Very special thanks are owed to P. Mick Richardson, Manager of Graduate Studies at the Missouri Botanical Garden, for his constant support and help.

LITERATURE CITED

Almeda, F. 1993. *Pilocosta* (Melastomataceae) revisited: A new species, polyploidy, and the base chromosome number of the genus. Novon 3: 311–316.

______. 1997a. Cytological and nomenclatural notes on the Mesoamerican species of *Aciotis* (Melastomataceae). Novon 7: 333–337.

______. 1997b. Chromosome numbers and their evolutionary significance in some neotropical and paleotropical Melastomataceae. BioLlania Edición Esp. 6: 167–190.

______. 2001. Melastomataceae. In *Flora de Nicaragua*, ed. W. D. Stevens, C. Ulloa Ulloa, A. Pool, and O. M. Montiel. Monogr. Syst. Bot. Missouri Botanical Garden 85: 1339–1419.

Almeda, F., and T. I. Chuang. 1992. Chromosome numbers and their systematic significance in some Mexican Melastomataceae. Syst. Bot. 17: 583–593.

Alvarez López, E. 1954 ["1953"]. Algunos aspectos de la obra de Ruiz y Pavón. Anal. Inst. bot. Cavanilles 12: 5–111.

Anderberg, A. 2001. Linnean Herbarium (S-LINN), Department of Phanerogamic Botany, Swedish Museum of Natural History (S). http://linnaeus.nm.se/botany/fbo/welcome.html.en

Aublet, J. B. C. F. 1775. *Histoire des plantes de la Guiane françoise*. London, Paris.

Baumgratz, J. F. A. 1983–1985. Morfologia dos frutos e sementes de Melastomataceae Brasileiras. Arch. Bot. Jard. Rio de Janeiro 27: 113–155.

Bentham, G. 1840. Enumeration of plants collected by Mr. Schomburgk in British Guiana. Melastomataceae. J. Bot. (Hooker) 2: 286–315.

Bonpland, A. 1818. *Monographie des Mélastomacées, Rhexiae,* 2: 93–104, pl. 35–40. Paris.

———. 1821. *Monographie des Mélastomacées, Rhexiae,* 2: 121–136, pl. 41–50. Paris.

Candolle, A. P. de. 1827. Memecyleae [Memecylaceae]. Linnaea 2: 505.

———. 1828. Melastomaceae. In *Prodromus systematis naturalis regni vegetabilis*, ed. A. P. de Candolle, 3: 99–202. Paris: Treuttel et Würtz.

Clausing, G., K. Meyer, and S. S. Renner. 2000. Correlations among fruit traits and evolution of different fruits within Melastomataceae. Bot. J. Linn. Soc., Bot. 133: 303–326.

Clausing, G., and S. S. Renner. 2001. Molecular phylogenetics of Melastomataceae and Memecylaceae: Implications for character evolution. Amer. J. Botany 88: 486–498.

Cogniaux, C. A. 1885. Melastomaceae: Tibouchineae. In *Flora brasiliensis*, founded by C. F. P. von Martius, ed. A. G. Eichler, 14(3): 206–510, pl. 108.

———. 1888. Melastomaceae. In *Flora brasiliensis*, founded by C. F. P. von Martius, ed. A. G. Eichler, 14(4): 397–656.

———. 1891. Melastomaceae. In *Monographiae phanerogamarum*, ed. A. de Candolle and C. de Candolle, 7: 1–1256. Paris: G. Masson.

Don, D. 1823. An illustration of the natural family of plants called Melastomaceae. Mem. Wern. Nat. Hist. Soc. 4: 276–329.

Freire-Fierro, A. 2001a. *Aciotis*. In *Flora of the Venezuelan Guyana*, ed. B. E. Berry, K. Yatskievych, and B. K. Holst, 6: 270–276.

———. 2001b. *Aciotis* (Melastomataceae): two new species from the northwestern lowlands of South America. Novon 11: 166–170.

Gleason, H. A. 1932. A synopsis of the Melastomataceae of British Guiana. Brittonia 1: 127–184.

———. 1935. Certain Melastomataceae of Suriname. Rec. Trav. Bot. Néerl. 32: 203–214.

———. 1958. Melastomataceae. In *Flora of Panama*, part 7(3), ed. R. E. Woodson et al., Ann. Missouri Bot. Gard. 45: 203–304.

Goldenberg, R., and G. J. Shepherd. 1998. Studies on the reproductive biology of Melastomataceae in "Cerrado" vegetation. Pl. Syst. Evol. 211: 13–29.

Görts-van Rijn, A. R. A., and M. J. Jansen-Jacobs. 1986. Melastomataceae. Flora of Suriname, Additions and Corrections 3(2): 356–429.

Greuter, W. 2001. Proposal to conserve the name *Trichostigma* against *Villamilla* (Phytolaccaceae). Taxon 50: 933–935.

Holmgren, P. K., N. H. Holmgren, and L. C. Barnett. 1990. Index Herbariorum. Part. I: The Herbaria of the world, 8th ed. Regnum Veg. 120: 1–693.

Krasser, F. 1893. Melastomataceae In *Die natürlichen Pflanzenfamilien*, ed. A. Engler and K. Prantl, III. 7: 130–199. Leipzig: Wilhelm Engelmann.

Lanjouw, J., and F. A. Stafleu. 1954. Collectors A–D. Index Herbariorum II, part 1. Regnum Veg. 2(2): 1–174.

Lemée, A. M. V. 1953. *Flore de la Guyana française*. Paris: Paul Lechévalier.

Linnaeus, C. [filius] 1782 ["1781"]. *Supplementum plantarum*. Braunschweig.

Macbride, J. F. 1941. Melastomataceae. In *Flora of Peru*, ed. J. F. Macbride, Publ. Field Mus. Nat. Hist, Bot. Ser, 13(4, 1): 249–523.

Mathies, P. S. 1981. A revision of *Aciotis* (Melastomataceae). Unpublished Ph. D. dissertation. Mississippi State University.

Metcalfe, C. R., and L. Chalk. 1972. *Anatomy of the dicotyledons: leaves, stem, and wood in relation to taxonomy, with notes on economic uses.* Oxford: Clarendon Press.

Miquel, F. 1840. Observationes des Piperaceis et Melastomaceis. In *Commentarii phytographici* by F. Miquel, 2: 76, pl. 10.

———. 1844. Symbolae ad floram surinamense. Linnaea 18(3): 272–301.

———. 1851. *Stirpes Surinamenses selectae*. Leiden: Arnz. & Soc.

Morley, T. 1976. Memecyleae. Flora Neotrop. Monogr. 15: 1–259.

Nationaal Herbarium Nederland (NHN). 1999. http://www.nationaalherbarium.nl:85/oopserver?ent=1&rec=300004&sct=50

Naudin, C. V. 1850. Melastomacearum monographicae descriptiones. Ann. Sci. Nat. Bot., sér. 3, 14: 118–165.

Ramírez, N., and B. Y. Brito 1990. Reproductive biology of a tropical palm swamp community in the Venezuelan llanos. Amer. J. Bot. 77:1260–1271.

Rao, T. A., and O. P. Bhupal. 1973. Typology of sclereids. Proc. Indian Acad. Sci. 77 B: 41–55.

Rao, T. A., and H. Jacques-Félix. 1978. Les types de sclérites foliaires et la classification des *Memecylon* africains. Adansonia, sér. 2, 18(1): 59–66.

Renner, S. S. 1984. Phänologie, Blütenbiologie und Rekombinationssysteme einiger zentralamazonischer Melastomataceen. Doctoral dissertation, University of Hamburg.

______. 1989. A survey of reproductive biology in neotropical Melastomataceae and Memecylaceae. Ann. Missouri Bot. Gard. 76: 496–518.

______. 1993. Phylogeny and classification of the Melastomataceae and Memecylaceae. Nord. J. Bot. 13: 519–540.

Ruiz, H., and A. J. Pavón. 1798–1802. *Flora peruviana, et chilensis*, 3 vols. Madrid: G. da Sancha.

______. 1954–1957. *Flora peruviana, et chilensis*, vol. 4. Anal. Inst. Bot. Cavanilles 12: 113–195, 13: 5–70, 14: 717–784, 15: 115–241.

Sell, Y., and G. Cremers. 1987. Les inflorescences des Melastomataceae guyanensis, leur filiation et leur valeur taxonomique. Canad. J. Bot. 65: 999–1010.

Skov, F. 1989. Hypertaxonomy—a new computer tool for revisional work. Taxon 38: 582–590.

______. 1990. Hypertaxonomy—a computer tool for revisional work. AAU Reports 26. Aarhus: Aarhus University Press.

Solereder, H. 1908. *Systematic anatomy of the dicotyledons: a handbook for laboratories of pure and applied botany*. Translated by L. A. Boodle and F. E. Fritsch; revised by D. H. Scott. Oxford: Clarendon Press.

Solt, M. L., and J. J. Wurdack.1980. Chromosome numbers in the Melastomataceae. Phytologia 47: 199–220.

Stafleu, F. A., and R. S. Cowan. 1981. Taxonomic Literature, ed. 2, vol. 3: Lh–O. Regnum Veg. 105: 1–980.

______. 1983. Taxonomic Literature, ed. 2, vol. 4: P–Sak. Regnum Veg. 110: 1–1214.

Stehlé, H., M. Stehlé, and L. Quentin. 1949. Catalogue des phanérogames et fougères. In *Flore de la Guadeloupe et dépendances* by H. Stehlé, M. Stehlé, and L. Quentin, 2(3): 101.

Standley, P. C. 1938. Melastomaceae. In *Flora of Costa Rica*. Publ. Field Mus. Nat. Hist, Bot. Ser. 18(3): 783–845.

Standley, P. C., and L. O. Williams. 1963. Melastomaceae. In *Flora of Guatemala*. Fieldiana: Bot. 24(7, 4): 407–525.

Stearn, W. T. 1992. *Botanical Latin*. 4th ed. Portland, Oregon: Timber Press,

Tieghem, M. P. van. 1891. Sur la structure et les affinités des Mémécylées. Ann. Sci. Nat. Bot. 13(2): 23–92, pl. 2.

Triana, J. J. 1871. Les Mélastomacées. Trans. Linn. Soc. London 28: 1–188.

Vegter, I. H. 1983. Collectors N–R. Index Herbariorum II, part 5. Regnum Veg. 109: 575–803.

Weberling, F. 1988. The architecture of inflorescences in the Myrtales. Ann. Missouri Bot. Gard. 75: 226–310.

Whiffin, T., and A. S. Tomb. 1972. The systematic significance of seed morphology in the neotropical capsular-fruited Melastomataceae. Amer. J. Bot. 59: 411–422.

Wurdack, J. J. 1962. Melastomataceae of Santa Catarina. Sellowia 14: 109–217.

______. 1970. Erroneous data in Glaziou collections of Melastomataceae. Taxon 19: 911–913.

______. 1973. Melastomataceae (Memecyleae by T. Morley). In *Flora de Venezuela*, ed. T. Lasser, 8: 1–819. Caracas: Instituto Botánico.

______. 1980. Melastomataceae. In *Flora of Ecuador*, ed. G. Harling and B. Sparre, 13: 1–406. Göteborg, Stockholm.

Wurdack, J. J., T. Morley, and S. Renner. 1993. 99. Melastomataceae. *Flora of the Guianas* 13: 1–301. Koenigstein: Koeltz Scientific Books.

APPENDIX 1

Chromosome Numbers

A. polystachya: n = 10, Peru (Solt & Wurdack 1980), Ecuador (pers. obs., *Freire-Fierro 2835*, MO, QCA); n = 20, Ecuador (pers. obs., *Freire-Fierro 2798*, MO, QCA).

A. acuminifolia: n = >10 (?), Ecuador (pers. obs., *Freire-Fierro 2805*, MO, QCA).

A. indecora: n = 26, Costa Rica (Solt & Wurdack 1980, *A. laxa*).

A. rubricaulis: n = 30, Venezuela (Solt & Wurdack 1980, as *A.* aff. *aristellata*).

A. indecora: n = 30, Costa Rica (Solt & Wurdack 1980, as *A. rostellata*), Panama (Almeda 1997b, as *A. rostellata*).

A. rubricaulis: n = 30, Costa Rica, Panama (Almeda 1997b, as *A. alata*).

A. purpurascens: n = ca. 40, Colombia (Solt & Wurdack 1980).

APPENDIX 2

Putative Hybrids

A. acuminifolia × *A. polystachya*

The Ecuadorian specimen *Harling 3246* (S), collected ca. 2 km from Puyo in Pastaza province, seems to be an hybrid. The specimen has the filiform sclereids and glandular leaf pubescence of *A. polystachya*, and the three-veined leaves and glabrous petals of *A. acuminifolia*. The inflorescence does not have any apparent developing fruits. The putative parental species have been collected in Puyo and its surroundings.

A. indecora × *A. purpurascens*

A collection from French Guiana (*Oldeman B-2093*) has the glabrous leaves and petals typical of *Aciotis purpurascens*, but the glandular trichomes on the stems, rachides, and hypanthia that are typical of *A. indecora*. Both putative parental species have been collected at the same locality (Bassin de l'Approuague, Arataye). Most of the other French Guianan collections seen of *A. purpurascens* have strigose petioles and leaves.

The Brazilian specimen *Daly 1130* also has characteristics of both *Aciotis indecora* and *A. purpurascens*. It has winged stems with eglandular and glandular hairs, densely glandular-pubescent leaves, glabrescent rachides, and glabrous fruits.

A. indecora × *A. rubricaulis*

The Colombian specimen *García et al. 67* (COL) from Quibdó, Chocó, has stems and petioles with long eglandular trichomes and the dorsal surface of the petals glabrous, like *A. rubricaulis*; however, the rachides and hypanthia are glandular-pubescent, characteristic of *A. indecora*. The inflorescences have only young buds and open flowers and lack any young fruits. Both putative parental species occur in the same locality.

A. polystachya × *A. rubricaulis*

Several Ecuadorian specimens collected in Puerto Quito (Pichincha) and its surroundings (*Freire-Fierro 2831*, *2832*, *2836* and *2838*) show the stoloniferous habit, elliptic to ovate and membranous leaves, and thyrsoid inflorescences that are typical characteristics of *A. rubricaulis*. The dense pubescence of these specimens is glandular, the ovary apex is crowned with glandular trichomes, and the anthers are dark purple, characteristics of *A. polystachya*. Additionally, the inflorescences show no mature fruits; immature fruits fall off very easily (pers. obs.). In such immature fruits examined, most of the seeds were undeveloped and whitish, unlike the typical brown seeds observed in young fruits of either likely parental species. Hybridization of *A. polystachya* with *A. rubricaulis* may be facilitated by the sympatric range of the parental species and visits by common pollinators; I observed the same Colletidae bees moving back and forth among the flowers of both putative parental species.

The collection *Fernández 210* (US) from Chocó, Colombia, has filiform sclereids in the leaves and hypanthia, like *A. polystachya;* however, it has the stems conspicuously winged, long eglandular trichomes, and glabrous petals, like A. *rubricaulis*. The inflorescences bear only buds and open flowers. A second collection similarly intermediate is *Freire-Fierro 2836* (QCA), from Puerto Quito, Ecuador. At the same locality, I collected both *A. rubricaulis* (*Freire-Fierro 2828*) and *A. polystachya* (*Freire-Fierro 2835*).

A. polystachya × *A. wurdackiana*

The Brazilian collection *Campbell P22514* (INPA, K, NY, US) from Pará was thought to represent a new species by Mathies and McDaniel (Mathies 1981). This collection along with *Campbell P22363* (GH, INPA, K, R, S, US) and *Campbell P22574* (F, GH, INPA, K, MG, NY, S), collected in Cachoeira Porteira, have, as in *A. polystachya*, lax inflorescences and rachides, and fruits covered by glandular trichomes. Like *A. wurdackiana*, they have very dense terminal paraclades, evident bracts, and whitish fruits. Three collections were obtained from the same locality where the putative parent species co-occur.

NUMERICAL LIST OF SPECIES

1. A. annua
2. A. wurdackiana
3. A. ornata
4. A. polystachya
5. A. ferreirana
6. A. paludosa
7. A. acuminifolia
8. A. olivieriana
9. A. circaeifolia
10. A. purpurascens
11. A. rubricaulis
12. A. indecora
13. A. viscida

INDEX TO NUMBERED COLLECTIONS EXAMINED

The numbers in parentheses refer to the corresponding species in the text and in the Numerical List of Species presented above. Collections of putative hybrids are noted by the symbol "×."

Acevedo-Rodríguez, P., et al. 3380 (10), 5937 (10).
Acosta Solís, M. 12157 (11), 12663 (11), 13939 (7).
Adams, C. D. 14204 (7), 14615 (10).
Adams, C. D., & Kalloo, M. B. 13966 (10).
Agostini, G. 1519 (7), 1567 (12).
Aguirre, L. E., et al. 520 (11).
Aguirre-Galviz, L. E. 962 (4).
Albert de Escobar, L. 39 (3), 40 (3), 41 (3), 45 (3), 301 (10).
Albuquerque, B., et al. 709 (12), 863 (10).
Alencar, L. 420 (13).
Allemão, F., & Cysneiros, M. de 626 (10).
Allen, C. 556 (7).
Allen, P. H. 218 (11).
Almeda, F. 541 (11).
Almeda, F., & Anderson, B. 5155 (12).
Almeda, F., & McPherson, G. 5962 (10).
Almeda, F., & Nakai, K. 3586 (12), 4031 (12), 4143 (12).
Almeda, F., et al. 3064 (10), 3089 (11), 3170 (12), 3195 (11), 3406 (11), 3408 (12), 3410 (11), 5866 (12), 5943 (12), 7491 (12), 7595 (12).
Alston, A. H. G. 485 (10), 5110 (10), 7640 (10), 7650 (8).
Alvarez, A., et al. 784 (4).
Alverson, W. S., et al. 177 (3).
Amaral, I. L., et al. 245 (7), 517 (10), 625 (4), 641 (10), 724 (12), 107109 (12).
Amorim, A. M., et al. 1375 (12).
Ancuash, E. 696 (8).
Andel, T. van, et al. 726 (10).
Anderson, A. 287 (1).
Anderson, R., & Mori, S. 32 (12), 172 (12).
Anderson, W. R. 10526 (7), 10621 (7), 12044 (7), 12205 (7), 13330 (7).
Andersson, N. J. 36? (6).
André, E. 3412 (10).
Antonio, T. 1199 (12), 1769 (11), 1813 (12), 1954 (10), 2020 (12), 2196 (11), 3231 (11), 4823 (11).
Aranda, J., et al. 1536 (10).
Araque, J., & Barkley, F. A. 18 M.034 (8), 18 V.169 (12).
Araquistain, M. 3019 (11).
Araquistain, M., & Moreno, P 2680 (11).
Araújo, A. P., & Ferreira, M. C. 440 (7).
Araújo, D., & Maciel, N. C. 4045 (6).
Arbeláez, A. 1 (10).
Arbeláez, M. V., & Sueroke, F. 749 (10), 761 (4).
Archer, W. A. 1432 (3), 1459 (10), 1668 (3), 1712 (12), 2036 (11), 2269 (1).
Arias, L. A., et al. 156 (11).
Aristeguieta, L. 3735 (10), 3983 (10).
Aristeguieta, L., & Lizot, J. 7384 (8).
Aristeguieta, L., et al. 6782 (10).
Arroyo, J. E. 19 (10).
Asplund, E. 4967 (12), 8546 (11), 8830 (11), 8848 (7), 8929 (10), 9107 (4), 9322 (10), 9349 (8), 13987 (10), 13988 (4), 14017 (7), 14067 (4), 14072 (10), 17278 (11), 17592 (4), 18316 (11), 18317 (10), 18327 (4), 18383 (11), 18663 (7), 19836 (8), 20105 (11).
Atkinson, D. J. 75 (12).
Atwood, J. T. 4818 (11), 5179 (11).
Aubréville, A. 251 (10).
Aulestia, C. 876 (12).
Austin, D. F. 4030 (1), 4069 (1).
Austin, D. F., et al. 7360 (10).
Ayala, F. 185 (10), 640 (4), 1889 (4), 2707 (7), 2874 (7), 2877 (10).
Ayala, F., & Arévalo, E. 4223 (7).
Ayala, F., et al. 2707 (7), 2851 (7), 2906 (11), 2924 (11), 3100 (7), 3100 (7), 3876 (4).
Aymard, G. 2188 (10), 4577 (10), 4719 (10).
Aymard, G., & Delgado, L. 6720 (10), 6854 (10), 8074 (12), 8211 (1).
Aymard, G., & Fernández, A. 7258 (10).
Aymard, G., & Stergios, B. 3326 (7).
Aymard, G., et al. 344 (12), 3514 (12).
Badcock, W. J. 360 (10).
Bahia, T. 88 (12), 148 (12).

Bailey, L., & Bailey, E. 581 (11).
Baker, C. F. 79 (9), 133 (1).
Balée, W. L. 1033 (10).
Balick, M., et al. 2607 (12).
Ballesteros, M. 13 (12), 35 (11).
Balslev, H., & Balseca, S. 4710 (11).
Bang, M. 2647 (11).
Barbour, P. J. 4928 (11).
Barclay, H. 4821 (11).
Barfod, A. 41016 (3), 41478 (11).
Barkley, F., & Bouthillette, G. 30C539 (12).
Barlow, F. D. 30/15 (12), 30/86A (12), 30/86B (12), 30/86 (12), 37/6 (12).
Barnard, 501 (10).
Barrier, S. 249 (10).
Barringer, K., et al. 2644 (11), 3745 (11).
Bartlett, A. W. 8299 (7).
Bautista, H. P. 136 (1).
Beck, H. T., et al. 128 (12).
Beck, S. G. 6527 (7), 6564 (11), 8071 (11), 13728A (11), 13728 (7), 17080 (10).
Beck, S. G., et al. 19416 (7).
Becket, J. E. 8535 (3).
Bedoya, N., et al. 44 (12).
Belém, R. P. 1639 (6), 1808 (12), 3482 (12).
Belém, R. P., & Aguiar, A. M. 1394 (11).
Belshaw, C. M. 3294 (10).
Bennett, B., & Gómez, P. 3519 (10).
Benson, W. 10263 (10), 10333 (10).
Berg, C. C., et al. P18575 (12).
Berg, M. E. van der 21 (12).
Berlin, B. 277 (10), 533 (10), 608 (10), 691 (10), 1506 (10).
Bernal, H. Y., & Téllez, H. 406 (10), 416 (4).
Bernal, R., et al. 2019 (11).
Bernardi, L. 2630 (1), 2719 (10), 2911 (12), 7108 (10).
Berry, P. 1437 (12).
Berti, L. M., et al. 1283 (10).
Betancourt, N. 152 (11), 200 (4), 217 (11).
Betancur, J. 1422 (10).
Billiet, F., & Jadin, B. 1639 (10), 4419 (10), 5927 (11).
Biolley, P. 6912 (11), 7446 (11), 7449 (12).
Bisby, F., et al. P18110 (3).
Bitaillon, C. 103 (10), 145 (12).
Black, G. A. 47-961 (9), 47-1824 (4), 48-2895 (9), 48-2977 (12), 48-3552 (12), 48-3572 (12), 49-8225 (10), 49-8405 (11).
Black, G. A., & Lobato, I. 50-9731 (10), 50-9699 (12).
Black, G. A., et al. 57-19543 (3).
Blackwell, W. H., et al. 2735 (12), 2744 (11).
Blade, A., & Burkhart, R. W. 3 (10), 5 (10).
Blake, S. F. 7509 (12).
Blanchet, M. 208 (12), 308 (6), 644 (6), 1030 (12), 1057 (11), 3951 (6).
Blanco, C. 162 (10).
Blum, K. E. 1326 (11).
Blum, K. E., et al. 2523 (12).
Boche, D. 19 (10).
Boeke, J. 3281 (10).
Boerboom, J. 9172 (12).
Bond, F. E., et al. 180 (7).
Bonpland, A. 49 (6).
Boom, B. M. 2549 (11).
Boom, B. M., & Gopaul, D. 7309 (10).
Boom, B. M., & Grillo, M. 6356 (10).
Boom, B., & Samuels, J. 8866 (10).
Boom, B. M., & Weitzman, A. L. 5933 (7).
Boom, B. M., et al. 8709 (12).
Bosbeheer, L. 10861 (12).
Bowie, J., & Cunningham, A. 47 (6).
Brade, A. C. 5037 (6), 8166 (6), 9088 (6), 19749 (6).
Brade, A. C., et al. 18364 (6).
Braga, P. I. S., et al. 3147 (4).
Brandbyge, J. et al., 30201 (10).
Brandbyge, J., & Asanza, E. 30804 (10), 32036 (10).
Brant, A., & Escobar, O. 1243 (10).
Bristol. M. L. 660 (7).
Britton, D. U. L., & Freeman, W. G. 9090 (10).
Britton, N. L., & Britton, E. G. 2092 (10).
Britton, N. L., & Hazen, T. E. 370 (10).
Britton, N. L., et al. 277 (10), 1253 (7), 1757 (10), 2532 (7).
Bro. Daniel 1615 (12).
Broadway, W. E. 2651 (10), 5088 (7), 5521 (7), 5645 (10), 5884 (10), 6024 (7), 6035 (10), 6519 (7), 6521 (10), 7293 (10), 7349 (10), 9147 (10), 9267 (7), 9807 (10).
Buchtien, O. 1052 (11), 1062 (11), 1710 (11), 1719 (11), 1733 (11), 2909 (11), 5473 (11), 5479 (11).
Budelman, A. 1041 (3).
Bunting, G. S. 3057 (11).
Bunting, G. S., & Alfonzo, L. 13274 (10).
Bunting, G. S., & Licht, L. 1245 (11).
Bunting, G. S., et al. 3460 (7), 3654 (12), 3793 (10), 3952 (1), 10842 (10), 10865 (10).
Burchell, W. J. 2627 (6), 9415 (1).
Burger, W., & Burger, M. 8114 (11).
Burger, W., & Ramírez, W. 3962 (11).
Burger, W., & Stolze, R. 4948 (11).
Burger, W., et al. 11673 (11), 12229 (11).
Buschbacher, R. J. 11 (1).
Busey, P. 378 (12), 821 (11).
Caballero, R. 166 (10).
Cabrera, I. 6307 (12).
Cadima, M. 51 (11).
Calderón, C., & Soderstrom, T. R. 2313 (4).
Calderón, C., et al. 2614 (3), 2962 (13).
Callejas, R. 449 (10), 483 (3), 1080 (3).
Callejas, R., & Atehortua, L. 318 (3).
Callejas, R., & Fonnegra, R. 7235 (10).
Callejas, R., et al. 1766 (12), 1998 (10), 2327 (12), 4118 (10).
Calzada, J. 5971 (12).

Camp, W. H. E-613 (11), E-1035 (10), E-1527 (10), E-2381 (11).
Campbell, D. G., et al. P20781 (10), P20793 (12), P20894 (12), P22085 (12), P22159 (9), P22363 (4×?), P22492 (8), P22514 (4×?), P22526 (2), P22572 (12), P22574 (4×?).
Campos, M. T., & Boom, B. 18 (10).
Cañas, G. 18 (10).
Cárdenas, D., et al. 6707 (7), 7097 (10).
Cárdenas, L. 2084 (11).
Cárdenas, M. 1898 (11).
Cárdenas, M., & Cutler, H. 7340 (11).
Cardona, F. 2778 (10).
Cario, B. 2895 (12).
Carleton, M. A. 461 (11), 487 (12), 492 (11).
Carmo, N. 185 (6).
Carpenter, J. B. 372 (11).
Carvalho, A. M. de, et al. 1073 (1), 4602 (12), 6075 (12).
Castillo, A. 776 (7).
Castro, E. 278 (11).
Cavalcante, P. 2445 (11).
Cazalet, P., & Pennington, T. 7755 (10), 5225 (7).
Cerón, C. 937 (10).
Chacón, A. 3 (11).
Chacón, I. 524 (12).
Chagas, J. INPA 2198 (5), INPA 50 (12), 33 (9).
Chaparro, M. L., et al. 16 (11).
Chavarría, M. M., et al. 23 (11).
Chávez, R. 1191 (11).
Chrysler, M. A., & Roever, W. E. 5283 (12).
Churchill, H.W., & Nevers, G. de 4254 (11).
Cid Ferreira, C. A., et al. 5238 (10).
Cid, C. A., & Lima, J. 3253 (12), 3711 (10), 3822 (12).
Cid, C. A., & Ramos, J. 1038 (2).
Cid, C. A., et al. 603 (9), 891 (3), 2053 (3).
Clark, H. L. 6978 (12), 6995 (12), 7013 (12).
Clewell, A. 3649 (12).
Coêlho, D. 1357 (10).
Coêlho, D., & Ramos, J. 50 (10).
Cogollo, A., & Borja, R. 530 (10), 540 (10).
Cogollo, A., & Cárdenas, D. 3725 (10).
Conant, D. S., et al. 1030 (12).
Contreras, E. 6543 (12), 6545 (12), 8952 (12), 11124 (11).
Contreras, N., & Agudelo, A. 17 (10).
Cooper, J. J. 10235 (11), 10949 (11).
Coradin, L., & Cordeiro, M. dos R. 1081 (9).
Cordeiro, M. 139 (10), 842 (12).
Cordero, G. 198 (11).
Córdoba, C., et al. 9 (12), 45 (12), 53 (12).
Correa, M. 331 (11).
Correa, M., et al. 3381 (11), 3856 (11).
Corrêa, M. P. F., et al. 101 (12).
Couret, 70 (10), 79 (12).
Cowan, C. 3328 (12).
Cowan, C., et al. 2299 (12).
Cowan, R. S. 2414 (10), 38203 (12), 38371 (10), 38375 (12), 38445 (11), 38461 (9), 38510 (10), 38911 (10), 38962 (10), 39226 (12).
Cowan, R. S., & Maguire, B. 38057 (12).
Cowan, R. S., & Richardson, D. 1169 (10).
Cowell, J. F. 139 (11).
Cremers, G. 4464 (10), 5242 (10), 6297 (10), 6402 (12), 6927 (10), 6928 (12), 6929 (3), 6948 (12), 7217 (10), 7237 (10), 7252 (12), 7292 (10), 7592 (10), 7919 (3), 7963 (10), 8105 (10), 8217 (3), 8381 (10), 10038 (10), 10893 (12), 12305 (12).
Cremers, G., & Granville, J. -J. de 13715 (10).
Cremers, G., et al. 10946 (10), 13533 (12).
Croat, T. 54107A (1), 7511 (4), 9893 (12), 9956 (12), 9993 (12), 11363 (12), 11580 (11), 12061 (12), 13012 (11), 13179 (12), 13318 (12), 17510 (13), 17519 (13), 17567 (13), 17829 (7), 18126 (4), 18325 (10), 18426 (10), 19068 (10), 19078 (10), 19961 (10), 24215 (12), 25452 (11), 27255 (11), 27570 (11), 33719 (12), 34012 (11), 34285 (12), 35806 (11), 36905 (12), 37648 (10), 37969 (12), 38183 (11), 44614 (11), 46959 (11), 51198 (10), 51747 (8), 53812 (12), 54134 (10), 54207 (12), 55939 (12), 56189 (12), 56305 (3), 56843 (8), 57504 (3), 59584 (1), 62211 (9), 62215 (9), 66890 (11), 70089 (12).
Croat, T., & Cogollo, A. 52151 (12), 52299 (3).
Croat, T., & Grayum, M. 59867 (11).
Croat, T., & Porter, D. M. 15295 (12).
Croat, T., & Rosas, A. 62436 (7).
Croat, T., & Watt, J. 70191 (4).
Croat, T., et al. 13086 (12).
Croizat, L. 360 (4), 702 (4), 955 (10).
Cuatrecasas, J. 1995a (7), 1995 (8), 3505 (8), 4515 (10), 4584 (8), 6792 (10), 6844 (12), 6985 (7), 7010 (12), 7417 (7), 7522 (10), 9050 (4), 10658 (10), 11451 (11), 13011 (10), 13012 (7), 14077 (11), 16109 (11), 16368 (12), 16954 (12), 19808 (12), 21417 (3).
Cuatrecasas, J., & Llano, M. 24089 (11).
Cueva, T. 8 (11).
Cufodonti, G. 558 (11).
Curiel, R., & Rodríguez, E. 155 (12).
Dahlgren, B., & Sella, E. 382 (1), 440 (1), 499 (1), 672 (1).
Daly, D. 1130 (10×?).
Daly, D., et al. D199 (1), D230 (10).
Daniel, R. 33 (10), 45 (10).
D'Arcy, W. G. 6759 (12), 9278 (11), 10191 (12), 10332 (11), 11133 (12), 14027 (10).
D'Arcy, W. G., & D'Arcy, J. J. 6699 (12).
D'Arcy, W. G., & Hamilton, C. 14968 (12).
D'Arcy, W. G., et al. 15529 (12).
Davidse, G. 2484 (10), 18280 (10).
Davidse, G., & González, A. C. 13097 (7), A14201 (7), 16401 (10), 21928 (10).

Davidse, G., & Miller, J. S. 26707 (1), 26929 (7), 27048 (7), 27283 (1).
Davidse, G., & Pohl, R. W. 1240 (11).
Davidse, G., et al. 1179 (11).
Davidson, C. 3438 (11).
Davidson, C., & Donahue, J. 8375 (12), 8702 (11).
Davidson, C., & Martinelli, G. 9987 (9), 10329 (4), 10364 (2).
Davidson, C., et al. 6714 (11).
Davis, P. H., & Coelho, D. F. 60311 (10).
Davis, P. H., et al. 59777 (6).
De la Cruz, J. S. 1021 (10), 1524 (10), 1573 (1), 1788 (1), 1790 (10), 1920 (10), 2083 (1), 2152 (1), 2155 (10), 2177 (10), 2468 (10), 2485 (10), 2486 (10), 2691 (12), 2776 (10), 2800 (12), 2827 (12), 2936 (10), 2971 (12), 2986 (10), 2987 (12), 3043 (10), 3115 (10), 3178 (12), 3181 (10), 3238 (10), 3270 (10), 3709 (10), 3790 (10), 3848 (1), 3990 (10), 3992 (1), 3999 (10), 4076 (10), 4455 (12).
Deam, C. C. 140 (11), 141 (12).
Del Carpio, C., et al. 1704 (7).
Delascio, F. 15370 (10).
Delascio, F., & López, R. 2767 (10), 8747 (10).
Delascio, F., & Reyes López 8847 (10).
Delascio, F., et al. 9324 (12).
Delgado, R. 98 (11).
Denslow, J. 218 (3), 2127 (12), 2161 (12), 2163 (12), 2242 (3), 2324 (3), 2405 (3), 2406 (3), 2407 (12).
Denslow, J., & Treacy, J. 35 (4).
Descoings, B., & Luu, C. 20374 (10), 20410 (10), 20540 (11).
Devia, W. 440 (12).
Devia, W., & Prado, F. 2540 (12).
Deward 11 (12), 80 (3), 193 (10).
Díaz, S. 3499 (11).
Dionysio 10807 (6).
Dodson, C. 5692 (7), 6283 (7).
Dodson, C. H., & McMahon, M. 4351 (7).
Dodson, C. H., & Torres, J. 2917 (10), 2957 (10).
Dodson, C., & Duke, J. 2233 (7).
Domínguez, G. 116 (12).
Don Jr., G. 110 (1).
Donnell Smith, J. 4791 (11), 6541 (11), 6553 (11), 6555 (11).
Dorr, L. J., et al. 4718 (10), 4875 (10), 7716 (10), 7794 (10).
Drouet, F. 2071 (1).
Dryander, E. 2609 (10).
Duarte, A. P. 6017 (6).
Duarte, A. P., & Gomes, J. C. 3736 (6).
Duke, J. A. MG-1101 (9), MG-17154 (1), 4467(a) (11), 4467 (11), 8628 (3) (12), 9407 (12), 9411 (11), 12059 (2) (12), 14794 (3) (12).
Duke, J. A., & Idrobo, J. 11519(3) (10).
Dunlap, V. C. 544 (11).
Duss, P. 4638 (7).
Dwyer, J. D. 7954 A (12), 1524 (12), 7088 (12), 7803 (12), 8407 (12), 8998 (12), 15153 (11).
Dwyer, J. D., & Correa, M. 8003 (11).
Dwyer, J. D., & Kirkbride, J. H. 7863 (12).
Dwyer, J. D., et al. 476 (12), 4703 (12).
Earle S., C., & Morgan Smith, H. 3439 (12).
Ebinger, J. E. 329 (12), 1116 (11).
Echeverry, R. 1659 (4), 2213 (10).
Eden, M. J. 44 (7).
Egler, W. 256 (1), 1401 (10).
Egler, W., & Irwin, H. 46439a (3), 46439 (3), 46452a (12).
Eiten, G., & Eiten, L. T. 10219-B (1).
Eiten, G., et al. 5203 B (9).
Elias, T. S. 270 (10).
Elias, T. S., & Hayden, S. M. 1806 (12).
Emmons, L. 3 (12).
Emygdio, L. 34 (6), 1134 (6).
Emygdio, L., & Emmerich, M. 3027 (6).
Emydgio, L., & Smith, L. B. 3086 (6).
Englesing, F. C. 272 (11).
Espina, J. 687 (3), 1809 (10).
Espina, J., & Arias, L. A. 1171 (4), 1180 (4), 1216 (4).
Espina, J., & García, F. 1922 (12).
Espina, J., et al. 1500 (3), 1501 (11), 1650 (3), 2464 (10), 3684 (10).
Espinosa, R. 5187 (11).
Estrella, M. R., & Bellusci, W. 31-15376 (1).
Eyerdam, W. J. 24674 (11), 24750 (11).
Fagerlind, F., & Wibom, G. 1113 (11), 1143 (11), 1721 (7), 1724 (7).
Falcão, M. 01 (9).
Fanshawe, D. 7 (10).
Fanshawe, D. B. 4812 (10), 5179 (3).
Farney, C., et al. 424 (10).
Fendler, A. 149 (10), 994 (10), 999 (7), 2268 (10).
Fernández Casas, & Susanna, A. 8362 (7).
Fernández, A. 398 (12), 1707 (1), 2678 (3), 3488 (3), 6055 (3).
Fernández, A., & Sanoja, E. 5111 (7).
Fernández A., et al. 10090 (11).
Fernández, J. L., et al. 7283 (11), 7463 (11), 8769 (4), 8901 (4), 11387 (4).
Fernández, R. 210 (4×?).
Ferreira, E. 6400 (7).
Ferreira, R. 16100 (7).
Ferreira, S. 94 (6).
Ferreyra, R. 4457 (10), 4899 (4), 5133 (13), 17257 (11), 17466 (10), 17572 (11), 18174 (11).
Feuillet, C. 304 (3), 1407 (3), 3623 (3).
Filskov, P., et al. 37074 (11).
Fleury, M. 145 (10), 233 (10), 590 (3).
Florschütz, J., & Florschütz, P. A. 1519 (10).
Focke, H. C. 584 (7).
Foldats, E. 2698 (10), 2729 (12), 2749 (12).
Folli, D. A. 1796 (6).
Folsom, J. P. 2546 (12), 9998 (11).

Folsom, J. P., & Butcher, H. 2800 (11).
Folsom, J. P., & Collins, L. 1528 (11).
Folsom, J. P., & Robinson, R. 2391 (12), 2397 (12).
Fonnegra, R., & Rengifo, W. 4838 (10).
Fonnegra, R., et al. 633 (3), 634 (3), 637 (12), 653 (3), 658 (12), 670 (12), 682 (12), 3133 (3), 4294 (10).
Forero, E., & Jaramillo, R. 2429 (11), 2430 (11), 2515 (3), 2532 (12), 5333 (12), 5376 (3).
Forero, E., et al. 944 (11), 1102 (12), 1148 (3), 1301 (3), 1486 (12), 3446 (11), 3811 (11), 3827 (12), 3828 (11), 3844 (12), 3885 (12), 4024 (11), 4189 (12), 4364 (12), 4555 (3), 4566 (12), 4567 (12), 4631 (9), 4782 (12), 4993 (3), 5206 (3), 5211 (3), 5217 (3), 5244 (7), 5440 (11), 5864 (11), 6081 (3), 6622 (11), 6948 (10), 6967 (11), 7177 (11), 8904 (3), 8938 (3), 9511 (3), 9546 (11), 9685 (3).
Forero, L. E. 728 (11).
Forest Dept. Br. Guiana 3283 (10).
Fosberg, F. R. 20161 (10), 29067 (7).
Foster, M. B., & Foster, R. 1678 (10).
Foster, R. 8751 (10), 13725 (7).
Foster, R., et al. 3107 (11), 11588 (11), 13832 (7).
Francisco INPA 1318 (12).
Frankie, G. W. 22a (11), 22c (11).
Freire-Fierro, A. 57 (11), 2767a (11), 2767b (11), 2768b (11), 2769a (11), 2769b (4), 2770 (11), 2775 (7), 2782 (10), 2783 (11), 2786 (10), 2787 (7), 2787b (7), 2788 (4), 2789 (7), 2790 (11), 2791 (11), 2793 (4), 2794 (10), 2795 (11), 2797 (7), 2798 (4), 2803 (7), 2805 (7), 2808 (11), 2810 (11), 2811 (4), 2815 (10), 2818 (11), 2827a (4), 2828 (11), 2829 (11), 2831 (4×?), 2832 (4×?), 2833 (11), 2834 (11), 2835 (4), 2836 (4×?), 2838 (4×?), 2838a (10), 2840 (10), 2842 (7), 2844 (4), 2989 (7).
Freire-Fierro, A., & León, S. 2744 (4), 2749 (10), 2750 (4), 2751 (10), 2755 (10), 2756 (4), 2759 (7), 2760 (11).
Freitas, M. A. 159 (13).
Friend, E. A. 29 (10).
Fróes, R. de Lemos 11518 (1), 20019 (6), 20031 (13).
Fromm, E., & Santos, E. 1256 (6), 1278 (6), 1353 (9).
Fromm, E., et al. 1410 (10), 1432 (10).
Fuchs, H. P., et al. 21818 (3).
Gabriel, M. 1108 (10).
Galeano, G., & Bernal, R. 407 (3).
Galeano, G., et al. 883 (10).
García, F. 34 (12).
García, L., et al. 49 (3), 67 (11×?), 79 (12).
García-Barriga, H. 7668A (12), 5014 (10), 7682 (12), 11165 (12), 13134 (11), 13135 (11), 13850 (13), 13854 (10), 15174 (3).
García-Barriga, H., & Lozano, G. 18279 (10).
Gardner, M. 932 (1), 999 (13), 1007 (11), 6017 (1).
Garwood, N., et al. 955 (11).
Gasche, J., & Desplats, J. 82 (10), 100 (1), 126 (12).
Geison, 485 (10).
Gentle, P. 1894 (11), 1916 (12), 4462 (12), 6063 (12), 7744 (12), 7907 (12), 8249 (11), 8249A (11), 8403 (11), 8601 (11), 9248 (11).
Gentry, A. 6862 (12), 15706 (7).
Gentry, A., & Berry, P. 14639 (7).
Gentry, A., & Brand, J. 36937 (3).
Gentry, A., & Daly, D. 18694 (11), 18706 (10), 18708 (11), 18266 (7).
Gentry, A., & Fallen, M. 17840 (3), 17896 (12).
Gentry, A., & Forero, E. 7205 (10).
Gentry, A., & Juncosa, A. 40582 (12).
Gentry, A., & Miller, J. 54973 (10).
Gentry, A., & Rentería, E. 23786 (3), 23859 (12), 24079 (12), 24159 (7), 79999 (10).
Gentry, A., & Revilla, J. 20769 (11).
Gentry, A., & Stein, B. 46928 (1).
Gentry, A., et al. 12761 (10), 15873 (4), 20758 (4), 21353 (10), 21996 (10), 22257 (10), 23662 (12), 25652 (10), 28922 (4), 30171 (3), 31600 (4), 35016 (11), 36405 (10), 36434 (10), 56712 (4).
Gillespie, L. J. 2893 (12).
Gillespie, L. J., & Persaud, H. 1600 (10).
Gillespie, L. J., & Tiwari, S. 2072 (3), 2119 (10).
Gillespie, L. J., et al. 2941 (10); 3013 (12).
Gillett, J. M., & Sampson, D. R. 16479 (10).
Ginzberger, A., et al. 603 (10).
Glaziou, A.-F.-M. 2570a (12), 569 (6), 2570 (6), 6151 (6), 9817 (1), 9822 (5), 9877 (4).
Gleason, H. A. 54 (10), 490 (10), 509 (1), 607 (10), 926 (10).
Godfrey, R. K. 66312 (11).
Gómez, L. D. 20019 (11).
Gómez, P., et al. 280 (10).
Gómez, R. 701 (7).
Gómez-Laurito, J. 10891 (12).
Gómez-Pompa, A., et al. 2939 (12), 3001 (12), 3187 (12), 3205 (12), 3307 (12).
González, M. S. 37 (3).
Gottsberger, G., & Döring, J. 19-12286 (7), 12-17286 (10), 12-18186 (10).
Gottsberger, G., & Morawetz, W. 12-31375 (9).
Gottsberger, G. K., & Posey, D. A. 117-20183 (10).
Graham, V. 170 (1).
Grández, G., & Vásquez, R. 625 (10).
Grant, J. R., et al. 91-01458 (11).
Granville, J. -J. de T.1056 (3), B.4364 (3), B.4887 (10), B.4976 (12), 15 (12), 612 (3), 1911 (10), 2535 (10), 2775 (10), 2911 (12), 2931 (12), 3155 (10), 3367 (12), 3451 (12), 3615 (11), 3767 (10), 4178 (10), 4190 (3), 4461 (10), 4462 (12), 5172 (10), 5346 (12), 5353 (12), 7038 (10), 7118 (12), 7613 (10), 9091 (12).
Granville, J. -J. de, & Cremers, G. 12815 (3), 12836 (10), 13059 (10).
Granville, J. -J. de, et al. 6053 (10), 6163 (12), 6319

(11), 10358 (12), 10624 (10), 10868 (10), 12705 (12).
Grayum, M. 1978 (12), 6669 (12).
Grayum, M., & Jacobs, B. 5347 (12).
Grenand, 200 (10).
Grijalva, A., & Moreno, P. 3463 (11).
Gröger, A., & Llamozas, S. 1124 (7).
Grubb, P. J., et al. 178 (7).
Guanchez, F. 1116 (10), 1254 (12), 1952 (12), 1957 (7).
Guanchez, F., & Melgueiro, E. 3641 (7).
Guanchez, F., & Mercado, L. 1915 (12).
Guarim Neto, G., & Neto, L. A. 528 (12).
Guedes, M. 2003 (9), 2004 (1).
Guedes, T. 142 (1).
Guillemin, M. 1037 (6).
Guizar, E. 507 (12).
Guppy, N. 7465 (3).
Gustafson, R. 1956 (7).
Gutiérrez, C. 35550 (12), 35585 (3).
Gutiérrez, G., & Schultes, R. E. 976 (12).
Gutiérrez, G., et al. 756 (10).
Hahn, W. 154 (12), 343 (12), 3596 (10), 3756 (10).
Hamblett, R. B. 295 (11), 451 (11).
Hamilton, C., & Krager, K. 3175 (12).
Hamilton, C., & Stockwell, H. 1425 (12).
Hammel, B. 772 (11), 1925 (11), 3656 (12), 4496 (12), 8159 (12), 11963 (12), 12182 (12), 12353 (11), 12436 (12), 12628 (11).
Hammel, B., & Trainer, J. 10762 (12), 12680 (12).
Hammel, B., et al. 19180 (12).
Hansen, B. 3018 (11).
Harley, R., et al. 10445 (7).
Harling, G. 964 (10), 985 (10), 3118 (11), 26062 (11).
Harling, G., & Andersson, L. 3246 (4×?), 11745 (10), 11946 (11), 12773 (11), 13047 (10), 13895 (11).
Harling, G., & Ståhl, B. 26749 (11).
Harling, G., et al. 19614a (7), 7187 (10), 7761 (10), 9324 (10), 15558 (11), 15579 (11), 19614 (10), 19642 (11).
Harmon, W. E., & Fuentes, J. A. 5151 (11).
Harrimanm, N. A. 17504 (10), 17587 (10), 17640 (7), 17699 (10).
Harris, S. A. TP 390 (10).
Harrison, S. G. 1208 (7).
Hatschbach, G. 26249 (6), 30948 (6).
Hatschbach, G., & Cervi, A. C. 51366 (6).
Hatschbach, G., & Kummrow, R. 38432 (1).
Hatschbach, G., & Silva, J. M. 50740 (6).
Hatschbach, G., et al. 58032 (6), 65213 (6).
Haught, O. 1314 (3), 1382 (10), 2090 (10), 2521 (7), 5574 (10).
Hayes, S. 165 (11).
Haynes, R. R. 7056 (12).
Hedger, J. 249 (12).
Heinrichs, E. 416 (8).
Hendrix, L. 203 (7), 218 (10).
Henkel, T. W., & Chin, M. 287 (10), 470 (10).
Heringer, E. P. 1726 (6).
Hermann, F. J. 11046 (7), 11133 (8).
Hernández, H., et al. 750 (11).
Hernández, J., et al. 116 (10), 537 (10).
Herrera, G. 1006 (11).
Herrera, H. 269 (10).
Hertel, H., & Oberwinkler, F. 15217 (7).
Hitchcock, A. S. 16937 (10), 16977 (1), 17323 (10), 17597 (3), 17653 (10).
Hoehne, F. C. 4262 (6), 5522 (7), 5531 (7).
Hoehne, F. C., & Gehrt, A. SP 42647 (6).
Hoehne, W. 15010 (6).
Hoff, M. 5416 (10), 5990 (10).
Hoff, M., et al. 6375 (3), 6553 (3).
Hoffman, B. 385 (7), 2938 (10).
Hoffman, B., & Aymard, G. 4004 (10).
Hoffman, B., & Capellaro, C. 859 (12), 901 (10), 925 (12), 1009 (7).
Hoffman, B., et al. 1344 (10), 1457 (12).
Hoffman, M., & Henkel, T. 3273 (10), 3801 (10).
Hoffman, W. A. 296 (6).
Holdridge, L. R. 6611 (12).
Holm, R., & Iltis, H. H. 340 (11), 943 (11).
Holm-Nielsen, L., et al. 2855 (7), 19604 (10), 19737 (4), 25293 (11), 25589 (11), 25725 (11).
Holst, B., & Liesner, R. 2658 (10).
Holst, B., & Werff, H. van der 2560 (10).
Holst, B., et al. 2110 (12).
Holt, E. G., & Blake, E. R. 421 (4), 428 (4), 585 (1).
Holt, E. G., & Gerhiger, W. 263 (7).
Hoover, S. 1055 (11).
Hoover, S., & Wormley, 1457 (11).
Hoover, S. A., et al 3797 (11), 4526 (11).
Hopkins, M. J. G. 24 (12).
Hostmann, F. W., 184 (7), 942 (7).
Hostmann, F. W., & Kappler, A. 31 (11), 215a (12), 215b (6), 1016 (12) 1832 (3).
Howard, R. A. 10324 (10), 10646 (7).
Hoyos, S., & Hernández, J. 412 (10), 597 (10).
Huber, J. MG-4187 (7), MG-4259 (11), MG-4975 (12), 114 (1), 672 (9), 866 (1), 1475 (10), 1497 (7).
Huber, O. 2582 (1), 12360 (10).
Hudson, J. 780 (7), 873 (11).
Huft, M., & Knapp, S. 1592 (11).
Hugh-Jones, D. Ll. 405 (11).
Humbert, H. 27398 (1), 27505 (1).
Hurtado, F., & Alvarado, A. 1091 (11).
Hurtado, F., et al. 1323 (10).
Hutchinson, P., et al. 6035 (11), 6049 (11).
Idrobo, J. M. 492 (7), 5273 (11), 5281 (3), 6788 (10), 8365 (8).
Idrobo, J. M., & Cuatrecasas, J. 2698 (3), 2706 (12).
Idrobo, J. M., & Schultes, R. E. 579 (8), 702 (10).
Induni, G. 269 (11).
Irwin, H. S. 165 (1).

Irwin, H. S., et al. 21228 (7), 47264 (10), 54919 (10), 55256 (7), 57538 (7).
Jangoux, J., & Bahia, R. P. 42 (1), 225 (1), 260 (10), 690 (7), 1098 (7).
Jangoux, J., & da Silva, M. F. F. 1208 (12).
Jansen-Jacobs, M. J., et al. 248 (10), 396 (10), 2481 (7), 2854 (10), 3096 (10), 3388 (7), 3392 (10), 4473 (10).
Janssen, A. 117 (9), 242 (1), 248 (7).
Janzen, D. H. 11643 (11).
Jaramillo, J. 2231 (10).
Jaramillo, J., & Coello, F. 4142 (10).
Jaramillo, J., & Rivera, S. 166 (10).
Jaramillo, J., et al. 31358 (10).
Jaramillo, R., et al. 7990 (10), 8090 (4).
Jaramillo-Mejía, R., et al. 1140 (7).
Jardim, A. 2449 (1).
Játiva, C., & Epling, C. 2018 (11), 2081 (11).
Jenman, G. S. 915 (1), 2094 (10), 2421 (12), 3631 (7), 3991 (9), 4013 (1), 4757 (7), 5430 (10).
Jeremie, J. 783 (7).
Jiménez, A. 743 (11), 1886 (11), 2118 (11), 4160 (12).
Johnson, W., & Barkley, F. 18C831 (10).
Johnson, W., & Homer, F. 271 (10).
Jones, G., et al. 3044 (12).
Jonker-Verhoef, A. M., & Jonker, F. P. 210 (10).
Jørgensen, P. M. 56427 (11).
Juajibioy, C. 233 (10).
Juncosa, A. 1137a (11), 991 (12), 993 (10), 1137 (3), 1334 (3).
Junk, W. 291 (7).
Kalloo, M. B. B957 (7).
Kappler, A. 1439 (10), 1520 (7), 1678 (10), 1680 (10).
Kayap, R. 27 (10), 874 (10), 1161 (10).
Keel, S., & Guedes, J. 316 (9).
Kellerman, W. A. 4509 (12), 7363 (12).
Kennedy, H. 1984 (11), 2634 (11), 2763 (12).
Kennedy, H., & Gra, S. 2229 (11).
Kennedy, H., et al. 2062 (12), 3014 (11).
Kernan, C. 164 (11).
Kessler, M., et al. 8233 (11), 8733 (11), 8830 (7), 9830 (11), 10784 (11), 11158 (11).
Killeen, T., et al. 7087 (12).
Killip, E. P. 5069 (12), 5071 (11), 5319 (12), 11520 (12), 26229 (10), 34255 (7), 34741 (10), 34922 (12), 35109 (3).
Killip, E. P., & García, H. 33125 (10), 33133 (11), 33257 (12), 33299 (11), 33300 (11).
Killip, E. P., & Smith, A. C. 14834 (10), 14925 (7), 25892 (11), 26888 (7), 26958 (10), 28173 (10), 28206 (10), 29852 (7), 30055 (9), 30085 (4, 30088 (7), 30367 (1).
Kimber, C. 723 (7).
King, G. 1188 (7).
King, R. M. 1116 (12).
King, R., & Guevara, A. 6116 (10).
King, S. 387-B (10), 387 (10).
Kirkbride, J. H., & Crebbs, T. C. 4 (11).
Kirkbride, J. H., & Lleras, E. 2797 (7).
Kirkbride, J. H., et al. 2703 (12).
Klug, G. 144 (10), 1188 (7), 1489 (10), 2174 (10), 3319 (11).
Knab-Vispo, C., & Rodríguez, G. 499 (10).
Knapp, S. 993 (11), 2671 (12).
Knapp, S., & Sytsma, K. 2446 (11).
Knapp, S., et al. 1687 (12).
Køie, 4761 (11).
Koster, L. B. 13032 (10).
Koyama, T., & Agostini, G. 7533 (12).
Kral, R. 25400 (12).
Kral, R., & Liesner, R. 71837 (10).
Kramer, K. U., & Hekking, W. H. 2233 (12), 2453 (10), 2780 (10), 3257 (12).
Krukoff, B. A. 10786 (11).
Krukoff, B. A., et al. 5889a (1), 5895a (12), 5074 (10), 6456 (12), 6971 (10), 7292 (12).
Kubitzki, K., & Poppendieck, H.-H. 79-22 (12).
Kuhlmann, J. G. 12 (7), 18 (9), 444 (5), 500 (4), 738 (6), 739 (7), 929 (7), 938 (9), 1122 (9), 1153 (7), 1154 (7), 1158 (10), 1215 (4), 1363 (10), 1967 (7), 4264 (7), 6543 (6).
Kuntze, O. 815 (10), 833 (10).
Kvist, L. P., & Asanza, E. 40745 (11).
Kvist, L. P., et al. 209 (1), 48316 (3).
Lachlianth, 254 (10).
Landeman, C. 2212 (9).
Landsberge, van 19 (8).
Langsdorff & Riedel 404 (12).
Lanjouw, J., & Lindeman, J. C. 978 (12), 1721 (10), 2883 (10).
Larpin, D. 282 (12), 353 (12).
Larsen, T. 1648 (10).
Lasseigne, A. P22586 (10), 22586 (10).
Lasser, T. 1645 (10).
Lawesson, J. E., et al. 44243 (13), 43981 (11).
Lawrance, A. E. 561 (10).
Lechler, 2186 (11), 2404 (11).
Lehmann, F. C. 29 (10), 5488 (12).
Leitão F., H. F., et al. 34632 (6).
Leng, H. 51 (10).
León, J. 1774 (12).
Leprieur, M. 1265 (10).
Lescure, J. P. 1 (12), 264 (1).
Level, J. S. 52 (7).
Levy, P. 497 (11).
Lewis, W. H., et al. 2328 (11), 2534 (12), 2813 (12), 3400 (11), 5476 (11), 9990 (7), 13509 (7).
Lewton, F. L. 435 (12).
Liesner, R. 1095 (12), 1868 (11), 3104 (11), 3490 (10), 3494 (12), 3596 (13), 6184 (12), 6201 (1), 6344 (10), 6797 (10), 7536 (10), 7545 (1), 8495 (13), 8841 (12), 17032 (12), 17267 (1), 19238 (10), 23508 (10), 23874 (10), 24296 (10), 24537 (4), 25656 (12).

Liesner, R., & Carnevali, G. 22762 (1).
Liesner, R., & Delascio, F. 21870A (13).
Liesner, R., & González, A. 9286 (4), 9305 (10), 10372 (10), 10784 (10).
Liesner, R., & Holst, B. 20107 (10), 20510 (12).
Liesner, R., & Medina, V. 13742 (10).
Liesner, R., & Morillo, G. 13799 (10), 13979 (10).
Liesner, R., et al. 15196 (11), 15448 (11), 20909 (10), 21902 (1).
Lima, A. 56-2608 (11).
Lima, L. de, et al. 567 (7).
Lindeman, J. 4220 (3), 5434 (10).
Lindeman, J., & Roon, A. C. de 879 (10).
Lindeman, J., et al. 10 (10), 170 (10), 399 (12).
Linden, J. 637 (12), 2268 (12).
Lisbôa, P. 137 (10), 865 (9).
Lleras, E., et al. P16895 (7), P17019 (12), P17029 (10).
Loaiza, C., et al. 118 (12).
Løjtnant, B., & Molau, U. 13318 (7), 13404 (7).
Lombardi, J. A. 1291 (6), 1328 (6).
Lombardi, J. A., & Temponi, L. G. 1099 (6).
López, A., et al. 4182 (10), 8693 (4).
López, R., & Giraldo-C., D. 602 (4), 690 (7).
López, R., et al. 822 (12).
López-Palacios, S. 2722 (10).
Lowe, J. 4017 (9), 4026 (9), 4208 (9).
Lowe, J., et al. 4120 (10).
Lowell, K., & Baker, 392 (11).
Lowrie, S. R., et al. 46 (12), 83 (10), 224 (7), 225 (7), 357 (7).
Lozano, G., & Lleras, E. 458 (7).
Lozano, G., & Ospina, M. 316 (10).
Lozano, G., & Rangel, O. 5316 (10), 5615 (11), 5628 (11), 5662 (10).
Lozano, G., & Ruiz, N. 7081 (11), 7082 (3).
Lozano, G., et al. 515 (10), 6111 (3), 6828 (11), 7157 (2) (12), 7398 (4) (10), 7437 (4).
Luetzelburg, P. 204 (11), 21110 (13), 21956 (9), 22110 (12), 22376 (7), 22489 (7), 22536 (12), 22560 (12), 22563 (12), 22696 (7), 22754 (12), 23998 (12).
Lugo, H. 187 (10), 1433 (7), 1447 (10), 1467 (7), 1484 (7), 1560 (10), 1696 (7), 2835 (10), 2935 (10), 3480 (10), 3850 (10), 4113 (10), 4146 (4), 4147 (10), 4923 (10), 4924 (11), 5529 (10), 5562 (10), 5591 (10), 5948 (10), 6038 (11), 6084 (11).
Lund 2283 (6).
Lundell, C. L., & Contreras, E. 19013 (12).
Luteyn, J. L. 962 (12), 1412 (12).
Luteyn, J. L., & Sylva, S. 13940 (11).
Luz, B. 1563 (6).
Lyra, R. P. de, et al. 155 (12).
Maas, P. J. M., & Maas, H. 215 (9).
Maas, P. J. M., et al. P13328 (7), 5453 (7), 5878 (10).
MacDougal, J. M., et al. 4180 (3).
Macedo, A. 3924 (7).
Macedo, M., & Assupes, S. 2959 (7).
Macía, M., & Asanza, M. 10 (10).
Macía, M., & Bensman, R. 49 (10), 55 (10).
Madison, M., et al. 25 (9), 111 (9).
Madrigal, B., et al. 753 (11).
Madriñán, S., & Barbosa, C. E. 233 (10), 428 (10).
Madsen, J. E., & Knudsen 86830 (11).
Maguire, B. 23739 (10), 24610 (11), 24761 (11), 24827 (10), 40500 (10).
Maguire, B., & Fanshawe, D. B. 23046 (1), 23566 (7).
Maguire, B., & Maguire, C. 44002 (12).
Maguire, B., & Politi, L. 27366 (13), 27367 (10).
Maguire, B., & Stahel, G. 25037 (12), 25046 (12).
Maguire, B., & Wurdack, J. J. 34625 (8), 34741 (12).
Maguire, B., et al. 23539 (7), 30359 (1), 30400 (1), 35842 (10), 36149 (10), 36425 (1), 36430 (10), 36536 (3), 36539 (7), 36699 (12), 36755 (7), 41951 (12), 56054 (10).
Maia, L. A., et al. 420 (12), 760 (10).
Marín 1702 (12).
Marinho, L. R. 69 (10).
Markgraf, F. 10254 (6).
Martin, J. H. 3784 (12), 3799 (12).
Martinelli, G. 6807 (9), 6820 (3), 7001 (4), 7039 (2).
Martinelli, G., et al. 12313 (1).
Martínez, E., & Riviere, R. 2063 (12).
Martínez, E., et al. 23156 (12), 23626 (12).
Martius, C. F. P. von 400 (12), 490 (6), 491 (6), 596 (6), 1044 (11).
Marulanda, O. 1205 (10), 2010 (10).
Mason, H. L. 13941 (4).
Mathews, 1279 (11), 1280 (4).
Mathias, M. E., & Taylor, D. 5425 (11).
Matuda, E. 3609 (12).
Mayor, E. 255 (1).
McDaniel, S. 10757 (10), 10758 (10), 13671 (10), 16180 (10), 19488 (7), 26336 (4).
McDaniel, S., & Persinos, G. 10694 (4).
McDaniel, S., & Rimachi, M. 16444 (10), 18439 (10), 24189 (7), 24759 (7), 25294 (7), 25313 (10).
McDaniel, S., et al. 27563 (10), 29837 (4), 29868 (10).
McDonagh, J. F., et al. 91 (12), 161 (11).
McDowell, T. 323 (12), 3404 (10).
McDowell, T., & Gopaul, D. 3083 (10), 3435 (10), 3519 (10).
McDowell, T., & Santana, M. 166 (12).
McDowell, T., & Stobey, A. 3872 (10), 3873 (10).
McDowell, T., et al. 4830 (10), 4941 (1).
McMahon, M. P. 4228 (7).
McPherson, G. 10703 (11), 11392 (11), 11503 (10), 12406 (11).
Medina, D. 10 (12).
Medri, M. E. 12 (9), 45 (9).
Meier, W. 240 (8).
Melin, D. 19 (7), 103 (10).
Menandro, M. S. 280 (6).

Méndez, R. 33 (12).
Mendonça, J., & Shima, D. 23 (9).
Mendonça, J., & Vieira, J. 28 (6).
Metcalf, R. D. 30664 (7).
Mexia, Y. 6039a (11), 6884 (10), 7201 (10), 7281 (10), 7329 (10), 8404 (11).
Meyer, G. 255 (7).
Meza, E. 505 (7), 519 (10), 559 (10).
Milliken, W. 533 (3).
Milliken, W., et al. 56 (10).
Miranda, F. E., et al. 922 (12).
Miranda, M.C., et al. 218 (8).
Mocquerys, A. 1045 (10).
Molina, A. 2070 (11), 14949 (12), 15028 (12).
Molina, A., et al. 17141 (11), 17634 (11).
Monsalve, M. 1021 (12).
Monteiro, O. P., & Damião, C. 199 (10).
Monteiro, O. P., & Ramos, J. 930 (10).
Monteiro, O. P., et al. 4 (10).
Moore, S. 471 (12), 730 (7).
Mora, L. E. 3973 (11), 4068 (11), 4146 (11), 4198 (11), 4200 (3).
Morales, J. F., et al. 1545 (12), 1939 (11).
Morawetz, W., & Wallnöfer, B. 112-241085 (10).
Moreno, P. 12361 (11), 13190 (11), 14801 (11).
Moreno, P., & Robleto, W. 20585 (11), 20669 (11).
Moreno, P., & Sandino, J. 12097 (11), 12788 (11), 12847 (11), 12875 (11), 12961 (11), 13013 (11), 13082 (11), 13120 (11), 14675 (12), 14914 (11), 15058 (11), 15117 (11).
Mori, S. A. 18497 (11).
Mori, S., & Anderson, R. 52 (11).
Mori, S., & Gracie, C. 18268 (11), 18382 (11), 19019 (7), 19028 (10), 21813 (10).
Mori, S., & Kallunki, J. 1811 (12), 2485 (12), 3476 (11).
Mori, S., & Thompson, C. B. 11035 (11).
Mori, S., et al. 8086 (10), 9071 (7), 10568 (6), 20362 (12), 20536 (10), 22043 (7), 22072 (10), 22332 (7).
Morillo, G. 2544 (10), 5487 (12).
Morillo, G., & Hasegawa, M. 5009 (1).
Morillo, G., & Ishikawa, M. 3480 (10).
Morillo, G., & Villa, N. 5452 (1).
Morillo, G., et al. 4075 (10).
Mosquera, E. 8 (3).
Muenscher, W. C. 12580 (11).
Murça, J., & Black, G. 1431 (9).
Nascimento, O. C. 369 (12), 586 (9).
Nee, M. 11282 (11), 31659 (11), 34879 (7), 41297 (7), 42504 (10), 42610 (10), 42689 (10), 42826 (10).
Nee, M., & Mori, S. 3644 (12).
Nee, M., & Vargas, I. 43248 (11).
Nee, M., & Vega, S. 27942 (11).
Neill, D. 1498 (11), 3342 (11), 3978 (11), 5880 (11).
Neill, D., & Vincelli, P. C. 3581 (11).
Nelson, B. W. 1048 (9).
Nelson, B. W., & Lima, J. F. P21124 (10).
Nelson, B. W., et al. 882 (9), 919 (10).
Nelson, C., & Romero, E. 4099 (12).
Nelson, E. B. 4118 (11), 5275 (11).
Nevers, G. de 3764 (11), 3898 (11).
Nevers, G. de, & Núñez, V. 3572 (12).
Nevers, G. de, et al. 3470 (12), 3522 (12), 4921 (12).
Noblick, L. R. 1784 (1).
Noblick, L. R., et al. 2405 (11).
Núñez, P. 6022 (11), 6582 (11).
Ochoa, M. 213 (12).
Oldeman, R. B-454 (3), B-455 (10), B-1003 (10), B-1559 (10), B-1767 (10), B-2093 (10×?), B-2191 (3), B-2637 (10), B-2654 (10), T-11 (3), T-718 (10), T-767 (7), T-834 (10), T-843 (10), 2275 (10), 2915 (12).
Oliveira, E. 6678 (1).
Øllgaard, B., & Molau, U. 13247 (10).
Øllgaard, B., et al. 35519 (11), 57243 (11), 90356 (11), 90409 (10), 90456B (11).
Opler, P. A. 94 (11), 525 (11).
Orcutt, C. R. 6514 (12).
Orozco, A. D. 147 (12).
Orozco, C., et al. 777 (12), 835 (7), 961 (12).
Osgood, W. H. 16 (7).
Ostenfeld, C. H. 83 (11).
Pabst, G. F. J. 9434 (9).
Padilla, I. 130 (11).
Palacios, P. A., & Jaramillo, R. 278 (4), 400 (3).
Palacios, P. A., et al. 405 (10).
Palacios, R. 4 (12).
Palacios, W. 3011 (10).
Palmer-Jones, R. W. 68 (10).
Passos, B. 1008 (10).
Paula, N. de, et al. 26 (10), 30 (9).
Peck, M. E. 870 (11).
Peixoto, A., et al. 3468 (6), 3512 (6).
Penland, C. W., & Summers, R. H. 225 (11).
Pennell, F. W. 1386 (10), 10977 (12), 10978 (10).
Pennell, F. W., & Killip, E. P. 5304 (12), 5985 (12).
Pereira, E. 3485a (12), 629 (7), 3341 (9), 3451 (9), 3485 (9).
Pereira, E., & Egler 3341a (12).
Pérez, L. 63 (12).
Perrottet, M. 699 (10).
Peterson, P. M., & Annable, C. R. 7197 (11), 7212 (11).
Philcox, D., & Freeman, B. 4741 (7).
Philcox, D., & Kalloo, C. 7210 (10).
Philipson, W. R., et al. 1401 (8).
Pinheiro, R. S. 1751 (6), 1846 (11).
Pinilla, A., & Gutiérrez, R. 14 (10).
Pinto, P., & Kotschwar, A. 101 (10).
Pinto, P., & Sastre, C. 1248 (7).
Pinto, P., et al. 339 (10).
Pipoly, J. J. 3538 (11), 3769 (11), 3917 (12), 3934 (12), 5034 (11).

Pipoly, J. J., & Boyan, R. 7591 (10), R. 8536 (12).
Pipoly, J. J., & Lall, H. 8110 (10), 8128 (10), 8213 (10), 8332 (3).
Pipoly, J., et al. 10794 (10), 10985 (10), 12110 (10), 12136 (10), 14954 (4).
Pires, J. M., & Black, G. A. 322 (1), 516 (1), 996 (7), 2423 (7).
Pires, J. M., & Cavalcante, P. B. 52582 (7).
Pires, J. M., & Silva N. T. 1541 (10), 1828 (10), 2426 (1), 4338 (1), 10482 (12).
Pires, J. M., et al. 4867 (9), 13871 (12), 50805 (10).
Pittier, H. 571 (11), 2440 (11), 2538 (11), 2572 (11), 3986 (11), 6516 (12), 9651 (12), 10293 (11), 10916 (10).
Pittier, H., & Durand, T. 9657 (12).
Pizziolo, W. 212 (6), 234 (6).
Plowman, T., & Schunke, J. 7539 (10).
Plowman, T., et al. 4202 (7), 8062 (12), 8917 (7), 12071 (7), 12558 (10).
Poeppig, E. 172 (10), 1837 (10), 1845 (12), 2496 (13), 2919 (7).
Poole, J. M. 1879 (1), 2032 (10).
Porter, D. M., et al. 4238 (12), 4377 (11), 4876 (12).
Potter, D. 5325 (10), 5395 (1).
Prance, G. T. 28034 (12), 28109 (10).
Prance, G. T., & Pennington, T. D. 1747 (7), 1959 (10), 1960 (10).
Prance, G. T., et al. 1586 (10), 2558 (3), 2876 (7), 3280 (7), 3335 (5), 3636 (9), 4268 (4), 4322 (10), 4899 (13), 5000 (12), 5130 (9), 5442 (10), 5506 (7), 5556 (11), 5745 (7), 6713 (11), 7425 (11), 7577 (7), 7615 (7), 8326 (12), 8684 (7), 8696 (7), 8706 (12), 9856 (10), 11172 (10), 11646 (9), 11839 (7), 12519 (4), 14346 (12), 15241 (7), 15686 (4), 16459 (10), 17992 (9), 22220 (2), 23586 (13), 23924 (7), 24400 (7), 25484 (3), 28747 (12), 30191 (12), 30454 (1).
Prescott, G. W. 451 (11), 903 (11).
Prévost, M. F. I-543 (3), I-765 (3), 1121 (12), 2065 (10).
Proctor, G. R. 17029 (7), 21540 (10), 21738 (7), 22442 (10), 32282 (11).
Proctor, G. R., et al. 27007 (11).
Pulle, A. 168 (10), 438 (7).
Purseglove, J. W. P.6451 (10).
Quesada, F. 240 (11).
Quiñones, L. 1577 (10), 2712 (10), 2742 (10).
Rabelo, B. V. 114 (10), 115 (10), 365 (10), 565 (12).
Rabelo, B. V., & Non, R. 1498 (10).
Rabelo, B. V., et al. 2655 (12), 3677 (12).
Ramírez, J.G., & Cárdenas, D. 680 (11), 1331 (11), 1516 (10), 1928 (10).
Ramírez, V. 188 (11).
Ramos, J. 1107 (3).
Raven, P. 21602 (11).
Renner, S. S. 203 (10), 214 (4), 363 (8), 753 (9), 918 (10), 919 (13), 955 (8), 2001 (4), 2105 (7), 2144 (12), 69323 (10), 69334 (8).
Rentería, E., et al. 1529 (10).
Reveal, J. L., & Balogh, P. 4954 (12), 4963 (11).
Revilla, J. 21 (7), 72 (10), 84 (10), 90 (10), 3526 (10), 3702 (10).
Reynel, C., et al. 5241a (11), 5241 (4).
Ribeiro, B. G. S. 1611 (12), 15187 (814) (9), 614 (10).
Rimachi, M. 395 (4), 448 (10), 595 (10), 645 (7), 2818 (7), 3969 (4), 4425 (4), 5576 (4), 5832 (10), 6294 (10), 7151 (10), 7266 (10), 7270 (4), 7323 (10), 7450 (10).
Ríos, D. 245 (11).
Rivero, R., et al. 1883 (10).
Rivière, R. 215 (11), 253 (12), 321 (12).
Robinson, H. E. 85-0060 (10).
Robles, R. 1881 (11).
Rodrigues, W. 485 (9), 10443 (7).
Rodrigues, W., & Chagas, J. 1950 (10), 1951 (10), 2148 (12), 2296 (12), 2628 (9), 3339 (10), 4140 (10), 4564 (12).
Rodrigues, W., & Coêlho, L. 1191 (9), 2278 (9).
Román, C. 72 (10).
Rombouts, H. E. 521 (11), 623 (10), 759 (10), 860 (10).
Romero, G. A., & Guanchez, F. 1800 (12).
Romero-Castañeda, R. 1213 (8), 1569 (10), 2709 (10), 3141 (10).
Rosa, N. A., & Vilar, H. 2831 (10).
Rosario, C. S., & Ubiratan, J. 738 (9).
Rovirosa, J. N. 888 (12).
Rowlee, W. W., & Stork, H. E. 702 (11).
Rudas, A., et al. 1380 (11), 2801 (7).
Rueda, R., et al. 1667 (12), 1853 (12), 2689 (11), 2737 (11), 3168 (11), 3191 (12), 3329 (11).
Ruiz, R., et al. 68 (10).
Ruiz-Terán, L. 3438 (10).
Ruiz-Terán, L., & López-Palacios, S. 10078 (1), 10957 (12), 10962 (1).
Ruiz-Terán, L., & Rodríguez, H. 6015 (10).
Ruiz-Terán, L., et al. 10810 (7).
Rusby, H. H. 2333 (11), 2582 (7).
Rusby, H. H., & Pennell, F. W. 500A (12).
Rusby, H. H., & Squires, R. W. 10 (7), 40 (7).
Rylands, A., & Hage, J. L. 178 (12).
Sagástegui, A. 5908 (10).
Sagástegui, A., & Aldave, A. 5797 (4).
Sagot, J. L. 212 (12), 214 (10).
Saldías, M., et al. 2702 (12).
Salick, J., & Quijano, J. 7659 (11).
Sampaio, A. J. de 4999 (10), 5111 (10).
Samuels, J. A. 109 (10), 268 (12).
Sanaiothi, T. M. 83 (1).
Sánchez, A. 24 (12).
Sánchez, P., et al. 507 (11).
Sandeman, C. 2277 (10), 3347 (10), 3666 (11), 5558 (12), 5571 (12), 5576 (12), 5577 (12).

Sandino, J. C. 1591 (11), 2722 (11), 3437 (11), 4660 (11).
Sandwith, N. Y. 56 (12), 545 (10), 1035 (7).
Sant'Alda, (6).
Santos, E. B. dos, & Alves, M. C. 137 (12).
Santos, E. B. dos, et al. 4251 (11).
Santos, J. U., et al. 35 (7).
Santos, T. S. dos 642 (12), 643 (6), 838 (11), 918 (6), 1127 (11), 1129 (6), 1338 (11), 1434 (6), 2197 (6), 2321 (11).
Sastre, C. 140 (12), 968 (8), 2165 (12), 2186 (1), 2260 (1), 3024 (12), 3096 (4), 3467 (1), 4527 (10), 5640 (10), 7402 (7).
Sastre, C., & Raichel, H. 4950 (4).
Sastre, C., & Sastre, F. 301 (10), 339 (12).
Saunders, J. 1205 (11).
Sauvain, M. 18 (10), 93 (10), 362 (3), 632 (10), 697 (10).
Saxicelli, 206 (12).
Schipp, W. A. 46 (12).
Schlim, L. 214 (10).
Schmidt-Munn, U. 420 (7).
Schnee, L. 1733 (7).
Schnell, C. 110 (11), 297 (12), 306 (12), 519 (11), 569 (11), 570 (11), 572 (11), 590 (12), 989 (11), 1006 (12), 1007 (11).
Schnell, R. 11380 (12), 11703 (3).
Schomburgk, R. H. 16 (1), 200 (12), 209 (1), 322 (7), 513 (7), 573 (7), 863 (9), 932 (1), 999 (13), 104/73 (10), 101/35 (10), 132/84 (10).
Schubert, B. G. 1116 (11), 2178 (9).
Schultes, R. E., & Black, G. A. 8393 (7).
Schultes, R. E., & Cabrera, I. 12521 (4), 13031 (9), 13164 (12), 13330 (7), 13743 (7), 15609 (7), 15775 (10), 16208 (10), 17204 (1), 17838 (10), 19435 (7), 19630 (10).
Schulz, J. P. 9327 (10).
Schunke, J. 146 (10), 213 (11), 816 (11), 1367 (11), 1498 (10), 1876 (11), 4851 (10), 4936 (10), 5557 (10), 6377 (10), 6553 (11), 6555 (11), 7345 (10), 7361 (10), 7854 (11), 8321 (11), 10285 (10), 10286 (10), 12223 (10), 13666 (11).
Schwabe, W. R. C. 67/335 (10).
Schwacke, C. 170 (1), 187 (9), 280 (1), 611 (1), 670 (1), 4148 (1), 9817 (1), 9822 (5), 9877 (4), 11980 (6).
Scolnik, R. 1161 (10).
Scolnik, R., & Luti, R. 573 (11).
Seemann, B. 31 (11), 1034 (10).
Sello, F. 47 (6); 888 (6).
Semple, K. S. 76 (11).
Sessé, M. 1202 (12).
Seymour, F. C. 3796 (11), 4136 (11), 5652 (11).
Shank, P. J., & Molina, A. 4215 (11), 4421 (11).
Shepherd, G. J. 10206 (6).
Shima, D., et al. 3 (10).
Shott 4148 (6).
Sidney, & Onishi, 1108 (6).
Signiera, R. MG-3771 (12).
Silva, A. 92 (1), 175 (9).
Silva, A. S. L. da, et al. 208 (10), 592 (12).
Silva, M. 512 (9), 548 (1), 913 (9), 1282 (7), 2123 (12).
Silva, M. F. da, et al. 309 (3).
Silva, M. F. da, et al. 38437 (1).
Silva, M. G., & Rosario, C. 4901 (10), 4922 (12).
Silva, M. N. 201 (7).
Silva, M., & Souza, R. 2221 (10), 2368 (10).
Silva, M., et al. 154 (12), 619 (12), 840 (12), 2342 (10).
Silva, N. T., 1559 (10), 2197 (10), 2435 (10).
Silva, N. T., & Santos, M. R. 4708 (2).
Silverstone, P. 557 (10).
Simpson, D. R., & Schunke, J. 641 (10).
Skov, F., et al. 7428 (10).
Skovsted, A. 47 (10), 93 (10).
Skutch, A. F. 3892 (12), 4520 (11), 4529 (7), 5521 (12), 5275 (11).
Smith, A. 3 (11), 3492 (10), 10079 (10).
Smith, D. 32 (12).
Smith, L. B. 1530 (6).
Smith, L. B., & McWilliams, E. L. 15407 (6).
Smith, S. F., et al. 679 (12), 739 (11), 740 (4), 1009 (11), 1103 (7), 1633 (12).
Sneidern, K. von A.36 (11), A.580 (11), A.1021 (4), 1198 (12), 4581 (12), 4582 (3), 4597 (12), 4966 (3), 5340 (12).
Snethlage, E. MG-9500 (7), 3 (7).
Snow, D. 32 (10).
Sobel, G., & Strudwick, J. 2374 (11).
Sodiro, A. 478 (11).
Soejarto, D. D. 2908 (3).
Soejarto, D. D., et al. 4423 (3).
Soleau, J. ACM 62 (12).
Solomon, J. 16927a (7), 3612 (10), 3613 (4), 7592 (12), 7687 (1), 16799 (12), 16927 (11).
Solomon, J., et al. 6584 (11), 6587 (11).
Sothers, C. A. 193 (9), 342 (9), 417 (8), 475 (9), 785 (9), 872 (10), 888 (12), 1004 (10), 1011 (8), 1093 (10).
Sothers, C. A., & Pereira, E. da C. 973 (9), 974 (9).
Sothers, C. A., & Silva, C. F. da 1005 (12).
Sothers, C. A., et al. 1012 (9).
Soukup, J. 3021 (10).
Souza, D. S., et al. 628 (6).
Souza, M. A. D. 359 (10), 360 (12), 1089 (12).
Sparre, B. 13620 (7), 14892 (11), 17593 (11), 17662 (11), 18031 (11), 18095 (11), 19045 (11), 19083 (11).
Sperling, C. R., et al. 5868 (10), 6172 (7).
Sprague, T. A. 104 (10).
Spruce, R. 141 (10), 337 (10), 338 (1), 529 (12), 654 (12), 710 (1), 816 (1), 977 (12), 1152 (7), 1153 (9), 1191 (7), 2123 (8), 2331 (13), 2451 (7).

St. George 310 (10), 573 (11).
Standley, P. C. 9106 (11), 24741 (12), 28620 (11), 30441 (11), 31123 (11), 36735 (11), 37023 (11), 72555 (12), 72663 (12), 72826 (12).
Standley, P. C., & Valerio, J. 47075 (11), 48840 (11), 51886 (11).
Steege, H. ter, et al. 308 (10).
Stehlé, H. 5103 (7).
Stehlé, H., & Stehlé, M. 5086 (7), 5588 (7), 6672 (7).
Steinbach, J. B. 7573 (11).
Steinbach, R. F. 451 (7), 458 (11).
Stelle, R. van 51 (12).
Stergios, B. 3213 (4), 3215 (7), 10352 (10), 10415 (12).
Stergios, B., & Aymard, G. 4413 (12), 7265 (8), 7392 (13), 7646 (12), 9235 (12), 9272 (12).
Stergios, B., & Delgado, L. 13763 (3).
Stergios, B., et al. 2737 (7), 3647 (10), 3955 (7), 6024 (10), 6241 (10), 7228 (7), 8102 (7), 8298 (12), 8914 (10), 13112 (1).
Stern, W., et al. 511 (10).
Stevens, W. D. 8014 (11), 8099 (11), 8640 (11), 12062 (11), 12392 (11), 12463 (11), 12525 (11), 18753 (11), 18860 (11), 19822 (12), 19966 (11), 20688 (11).
Stevens, W. D., & Montiel, O. M. 24663 (12).
Stevens, W. D., & Moreno, P. P. 19228 (11), 19626 (12).
Stevens, W. D., et al. 5013 (11), 19562 (12), 25060 (11), 25174 (12), 25615 (12).
Steward, W. C., et al. 11 (10), 12 (9), 33 (3).
Steyermark, J. A. 17417 (11), 38093 (12), 41855 (11), 42043 (12), 56729 (3), 58097 (1), 60444 (10), 60524 (12), 60533 (10), 74805 (12), 74835 (11), 75972 (12), 87182 (3), 87184 (10), 87704 (7), 89072 (10), 89390 (10), 89592 (12), 90283 (12), 90422 (12), 90533 (3), 90719 (7), 98226 (4), 99967 (10), 106168 (4), 124274 (3).
Steyermark, J. A., & Bunting, G, 102703 (8), 102792 (10).
Steyermark, J. A., & Delascio, F. 129053 (4).
Steyermark, J. A., & Fernández, J. E. 99607 (10).
Steyermark, J. A., & Gibson, H. 95663 (10), 95774 (12).
Steyermark, J. A., & Liesner, R. 119065 (11), 119219 (7).
Steyermark, J. A., & Luteyn, J. 129721 (7).
Steyermark, J. A., & Rabe, M. 96048 (10), 96523 (10), 96619 (7).
Steyermark, J. A., & Wurdack, J. J. 66 (10), 1331 (12).
Steyermark, J. A., et al. 92650 (12), 101319 (10), 107106 (3); 109874 (10), 113898 (10), 115160 (10), 117721 (12), 119426 (10), 119462 (10), 119490 (7), 119496 (10), 119592 (10), 122296 (10), 125671 (4), 125799 (1), 127966 (7).
Stockdale, F. A. 8807 (10).
Stoffers, A. L., et al. 157 (10).
Straw, R. M. 2414 (10).
Sucre, D. 230 (12), 5644 (6).
Sullivan, G. A. 50 (12), 641 (12).
Susach, F. 1032 (1).
Sytsma, K. 1266 (11), 1748 (12), 1766 (12), 4033 (12).
Sytsma, K., & D'Arcy, W. G. 3571 (11).
Sytsma, K., & Sytsma, R. 3103 (11), 3119 (11).
Tate, G. H. H. 2 (10), 34 (9), 77 (1), 93 (13), 103 (10), 467 (11), 973 (7).
Tate, R. 70 (11).
Taylor, J. 17956 (11).
Teixeira, L. O. A., et al. 903 (10), 969 (9).
Téllez, O., et al. 4886 (11), 5051 (11).
Tello, & Tina, F. 2180 (4).
Tessman, G. 3032 (7).
Thomas, R. 10 (3).
Thomas, W. W., et al. 3931 (7), 9243 (6), 10513 (12), 10805 (12).
Thompson, S. A. 4688 (12).
Thompson, S. A., & Rawlins, J. E. 841 (11).
Tillett, S. S. 676-410 (4), 738-563 (10), 763-99 (10).
Tillett, S. S., & Tillett, C. L. 45327 (10), 45742 (12).
Tina, F., & Oliveira, F. 2317 (4).
Tonduz, A. 8116 (11), 8118 (11), 8262 (11), 11443 (11), 13136 (11), 14842 (11).
Toriola, D. 1045 (12), 1046 (12).
Toro, R. A. 1122 (10), 1347 (12).
Torres, J. H., et al. 3069 (4).
Trail, J. W. H. 252a (1), 247 (7), 250 (9), 252 (1), 253 (1).
Treacy, J., & Alcorn, J. 20 (10), 66 (10), 226 (10), 262 (10), 416 (7), 525 (10).
Triana, J. J. 3880 (10), 3881 (12), 6195 (12), 6195a (8).
Tsugaru, S., & Sano, Y. B-630 (9).
Tuerckheim, N. 1218 (12), 8839 (12).
Tún Ortíz, R. 2073 (12), 2438 (11).
Tunqui, S. 458 (10), 459 (4).
Tutin, T. G. 5 (1), 477 (1), 514 (12), 517 (9), 533 (12), 534 (12).
Tyson, E. L., et al. 3308 (11), 4329 (11).
Ule, E. 2394 (6), 5087 (9), 5092 (12), 5096 (7), 5663 (11), 6232 (10), 6267 (4), 9667 (11).
Umaña, G., & Chacón, R. 422 (11).
Uribe, L. 24 (10), 2850a (10), 1305 (11), 1605 (11), 2850 (12), 2928 (12), 3050 (10), 3112 (4), 3129 (12), 3131 (12), 3136 (12), 3592 (10), 3929 (7), 3964 (10), 4407 (10), 4533 (7), 6579 (4).
Uribe, L., & Pérez-Arbeláez, E. 2715 (10).
Uribe, L., et al. 44a (12), 44 (3).
Utley, J., & Utley, K. 1218 (11), 2702 (11).
Vargas, C. 3786 (11), 13982 (11), 14571 (11), 16889 (7), 116661 (11).
Vargas, I. 564 (7).
Vargas, O. 99 (11).
Vargas, O., & Frazee, J. 21 (11).

Vargas, O., et al. 1656 (11).
Vasconcellos, M. B., & Carvalho, R. F. M. de 10329 (10).
Vásquez, R. 3957 (10), 13377 (7).
Vásquez, R., & Jaramillo, N. 462 (10), 974 (10), 1176 (10), 1182 (4), 3291 (10), 4893 (10), 14450 (10), 15935 (4).
Vásquez, R., et al. 6391 (7), 13763 (10).
Vaughan, G. 53 (7).
Velloso, E. P. 798 (12), 1096 (3).
Velloso, H. 1035 (6), 1039 (12), 1041 (12).
Vera, N. 52 (10).
Vergara, D., & Torres, R. 112 (12).
Vianna Freire, C. 56 (6), 789 (6).
Vieira, G., et al. 216 (7), 696 (10).
Vilhena, R., et al. 264 (12).
Vogel, S., et al. 341 (8).
Vrieze, J. 4351 (7).
Wachenheim, G. 33 (12), 104 (12), 114 (10), 141 (10).
Waide, R. 62301 (3), 62302 (3).
Wallnöfer, B. 24-19588 (10).
Warming, 163 (7).
Warner, R. H. 59 (10).
Warner, R. H., & White, J. W. 75 (3).
Watson, S. 392a (12).
Wawra, H., & Maly, 161 (10).
Wedel, H. von 582 (12), 782 (11), 1450 (11), 2863 (12).
Weitzman, A. 215 (10).
Werff, H. van der, & González, A. C. 4836 (10).
Werff, H. van der, & Ortega, F. 8110 (10).
Werling, L., & Leth-Nissen, S. 570 (7).
Wessels Boer, J. G. 552 (7), 653 (10), 963 (10), 2094 (10).
Whiffin, T. P. 11 (10), 33 (12).
White, G. E. 2091 (7), 2092 (11).
White, S., et al. 177 (3).
Whitefoord, C., & Eddy, A. 349 (10).
Whitton, B. A. 56 (1), 58 (10).
Wilbur, R. L., & Luteyn, J. L. 11617 (12), 11701 (12), 11840 (12).
Wilbur, R. L., & Stone, D. E. 10602 (11).
Wilbur, R., L. & Teeri, J. 13359 (11), 13635 (11).
Wilbur, R. L., & Weaver, R. E. 10842 (12), 11351 (11).
Wilbur, R. L., et al. 11149 (12), 15560 (11).
Williams, L. 38 (7), 334 (10), 360 (10), 406 (10), 463 (10), 803 (10), 883 (7), 1179 (12), 1180 (12), 1185 (7), 1276 (10), 1474 (10), 1485 (10), 1486 (7), 1688 (10), 1710 (7), 1712 (7), 1930 (10), 2568 (10), 2586 (10), 2701 (10), 2978 (4), 3209 (10), 3813 (10), 4739 (10), 6362 (4), 7493 (11), 7848 (10), 11673 (10), 12744 (7), 15779 (10).
Williams, L., & Sella, E. 39 (1).
Williams, L., et al. 28461 (12).
Williams, R. O. 10339 (10), 12870 (10).
Williams, R. S. 735a (10), 289 (12), 735 (12).
Witherspoon, J. T., & Dressler, R. 8917 (11).
Woodruff, L. 593 (10).
Wormley, S. 1586 (11).
Woronow, G., & Juzepczuk, S. 5887 (4).
Woytkowski, F. 6077 (4), 8182 (11), 34411 (11), 34412 (11), 35084 (10).
Wullschlägel, H. R. 22 (12), 785 (7), 786 (12), 1449 (7), 1759 (12), 2042 (12), 2043 (10), 2044 (7).
Wurdack, J. J. 1827 (10), 1848 (10), 2634 (10), 2851 (11).
Wurdack, J. J., & Adderley, L. S. 42904 (12), 43416 (1).
Wurdack, J. J., & Monachino, J. V. 41158 (10).
Wurdack, J. J., et al. 2847 (12), 2848 (11).
Young, K. 991 (11).
Zamudio, S., et al. 92 (12).
Zarucchi, J. 3278 (12).
Zarucchi, J., & Barbosa, C. 3812 (7).
Zarucchi, J., et al. 6736 (10).
Zimmermann, G. 37 (13).

INDEX TO SCIENTIFIC NAMES

Accepted names are in roman type; the main entry for each is in **boldface**. Synonyms are in *italics*.

Aciotis D. Don **18–19**
  sect. *Euaciotis* [= *Aciotis*] 3, 18
  sect. *Spennera* (Mart. ex DC.) Krasser 3, 18
  acuminifolia (Mart. ex DC.) Triana 2, 3, 4, 5, 6, 8, 9, 10, 11, 12, 13, 14, 18, 19, 23, 29, 32, **39–44,** 47, 59, 83, 84, 85
  acutiflora (Mart.) Triana 68, 77
    var. *parvifolia* Cogn. 69
  aequatorialis Cogn. 18, 29, 39, 44
  alata (Beurl.) Almeda 18, 60, 83
  amazonica Cogn. 39, 43, 44
    var. *radicans* Cogn. 39
  annua (Mart. ex DC.) Triana 2, 4, 5, 6, 8, 9, 10, 12, 13, 16, 18, 19, **20–24,** 33, 44, 85
  anomala Brade 69, 77
  aquatica (Aubl.) D. Don ex Loudon 80
  aristata Ule 60, 68
  aristellata Markgr. 60, 83
  asplundii Wurdack 60, 68
  brachybotrya (DC.) Triana 34
    var. *ferruginea* (Naudin) Cogn. 35
  caulialata (Ruiz & Pav.) Triana 60, 68
  circaeifolia (Bonpl.) Triana 2, 3, 4, 5, 6, 9, 11, 12, 14, 16, 18, 19, 33, 38, **47–50,** 85

var. *major* Cogn. 50, 60
cordata (Vell.) J. F. Macbr. 44, 80
dichotoma (Benth.) Cogn. 18, 39, 43, 44
var. *anomala* (Miq.) Cogn. 39, 43
var. *longifolia* S. Moore 39
discolor D. Don 3, 13, 18, 50
var. *martinicensis* (Naudin) Stehlé 50
dysophylla (Benth.) Triana 16, 20
ferreirana Brade 2, 4, 5, 6, 9, 11, 14, 16, 18, 19, **32–33,** 34, 35, 85
ferruginea (Naudin) Triana 35
fragilis (DC.) Cogn. 50
var. *lancifolia* (Steud. ex Cogn.) Cogn. 51
herbacea Cogn. 26
indecora (Bonpl.) Triana 2, 3, 4, 5, 6, 9, 11, 12, 13, 14, 17, 18, 19, 32, 47, 59, **68–77,** 78, 83, 84, 85
var. *glabrescens* Cogn. 77
var. *macrophylla* Cogn. 69, 77
var. *sagotiana* Cogn. 69
latifolia (DC.) Cogn. 34
laxa (Rich. ex DC.) Cogn. 68, 83
var. *kappleriana* (Naudin) Cogn. 69
var. *robusta* Cogn. 69, 77
levyana Cogn. 60, 68
longifolia (Mart. ex DC.) Triana 47, 50
var. *glabra* Huber 51
martiana Cogn. 35, 44
martinicensis (Naudin) Urb. 50
olivieriana Freire-Fierro 2, 4, 5, 6, 8, 9, 10, 18, 19, **44–47,** 85
ornata (Miq.) Gleason 2, 4, 5, 6, 8, 9, 10, 17, 18, 19, **26–29,** 43, 85
paludosa (Mart. ex DC.) Triana 2, 4, 5, 6, 9, 11, 14, 18, 19, **33–39,** 43, 85
var. *latifolia* (DC.) Triana 34
pellucida (DC.) Pittier et al. 50
pendulifolia (Bonpl.) Triana 80
polystachya (Bonpl.) Triana 2, 4, 5, 6, 8, 9, 10, 11, 12, 13, 14, 16, 18, 19, 22, 23, 24, **29–32,** 43, 47, 68, 78, 83, 84, 85
purpurascens (Aubl.) Triana 2, 3, 4, 5, 6, 8, 9, 10, 11, 12, 13, 14, 16, 18, 19, 38, 43, 48, **50–60,** 77, 83, 84, 85
var. *alata* (Beurl.) Cogn. 59, 60
var. *longifolia* Cogn. 51
var. *martinensis* J. F. Macbr. 60
var. *martinicensis* (Naudin) Stehlé, M. Stehlé & Quentin 50
var. *pellucida* (DC.) Cogn. 50
rostellata (Naudin) Triana 69, 77, 83
rubricaulis (Mart. ex DC.) Triana 2, 3, 4, 5, 6, 8, 9, 11, 12, 13, 14, 18, 19, 32, 44, 59, **60–68,** 77, 83, 84, 85
sieberi (Steud.) Triana 81
sileniflora (Bonpl.) Triana 50
sphaeranthera (Naudin) Cogn. 39, 43,
spiritusanctensis Brade 35, 39
trinitensis Cogn. 51
uliginosa (Bonpl. ex Naudin) Triana 50
viscida (Benth.) Freire-Fierro 2, 4, 5, 6, 8, 9, 10, 18, 19, 35, 47, **77–79,** 85
viscosa (Naudin) Triana 69
wurdackiana Freire-Fierro 2, 4, 5, 6, 8, 9, 10, 14, 18, 19, 23, **24–26,** 32, 33, 84, 85
Acisanthera P. Br. 2, 17
Apidae 44
Apinae 44
Appendicularia DC. 17
Arthrostema
caulialatum Ruiz & Pav. 3, 60, 68
Augochlora Smith 1853 32
callichroa 32
Bellucia Neck. ex Raf. 4
Brachyotum (DC.) Triana 12
Chaetolepis (DC.) Miq. 12
Clidemia
herbacea Steud. ex Cogn. 80
Colletidae 32
Comolia DC. 2, 17
Ernestia DC. 2, 17
Euglossa Latreille 1802 32
Exomalopsini 44
Henriettea DC. 4
Heterocentron Hook. & Arn. 12
Heterotis Benth. 12
Leandra Raddi 17
montana Gleason 17, 69
Macairea DC. 2
Megachilidae 32
Megommation Moure 1943 50
ogilvei 44
Melastoma L. 11
annuum Schrank ex DC. 80
aquatica Aubl. 80
cordata Vell. 80
fragilis Rich. ex DC. 80
laxa Rich. ex DC. 80
pellucens Rich. ex DC. 80
purpurascens Aubl. 3, 50, 59
purpurea Sw. 50
purpurea Willd. 50
rubricaule Schrank ex DC. 80
uliginosa Bonpl. ex Naudin 80
viscosa Bonpl. ex DC. 80
Melastomataceae 1, 4, 5, 10, 12, 17, 47, 50, 68, 77, 80
Melastomeae 1, 2, 12, 17
Melipona Illiger 1806 44
fulva 44
Memecylaceae DC. 4
Memecylon L. 4
ramiflorum 5
Miconia Ruiz & Pav. 3
purpurascens (Aubl.) DC. 50
Miconieae 4, 17
Microlepis (DC.) Miq. 2
Monochaetum (DC.) Naudin 12
Mouriri Aubl. 4
Myrtales 5
Nepsera Naudin 17
aquatica (Aubl.) Naudin 81
pendulifolia (Bonpl.) Naudin 80
Ossaea DC. 17
bracteata 18
robusta 18

Paratetrapedia Moure 1941 23, 32
duckei 44
Paratrigona Schwarz 1938 60
Pilocosta Almeda & Whiffin 12
Pterolepis (DC.) Miq. 12
Pternandra W. Jack 11
Rhexia L. 3
aquatica (Aubl.) Sw. 80
circaeifolia Bonpl. 50
herbacea Raddi 33, 38
hexandra Schrank ex DC. 80
indecora Bonpl. 68, 77
pendulifolia Bonpl. 80
polystachya Bonpl. 29
sileniflora Bonpl. 50, 59
Siphanthera Pohl 17
Sonerila Roxb. 4
Sonerileae Triana 4
Spennera Mart. ex DC. 3, 18
acuminifolia Mart. ex DC. 39
acutiflora Mart. 68, 77
alata Beurl. 59, 60
annua Mart. ex DC. 18, 20
anomala Miq. 39
aquatica (Aubl.) DC. 80
brachybotrya DC. 34, 38
caulialata (Ruiz & Pav.) Naudin 60
circaeifolia (Bonpl.) DC. 47
circaeoides Mart. & Schrank ex DC. 47
dichotoma Benth. 39, 44
var. *lanceolata* Naudin 39, 44
dysophylla Benth. 20
ferruginea Naudin 35, 38
fragilis DC. 50
grandifolia Miq. 50
indecora (Bonpl.) DC. 68
kappleriana Naudin 69
lancifolia Steud. ex Cogn. 81
latifolia Benth. 20, 34
latifolia DC. 20, 34
laxa Rich. ex DC. 68
longifolia Mart. ex DC. 47
martinicensis Naudin 50
ornata Miq. 26
paludosa Mart. ex DC. 33
panicularis Naudin 60, 68
pellucida DC. 50
var. *subcordata* Crueg. 50, 59
pendulifolia (Bonpl.) DC. 80
polystachya (Bonpl.) DC. 29
rostellata Naudin 69
rubricaulis Mart. ex DC. 60
sieberi Steud. 81
sileniflora (Bonpl.) DC. 50
sphaeranthera Naudin 39
tetraptera Miq. 20
uliginosa Bonpl. ex Naudin 50
viscida Benth. 77
viscosa Naudin 69
Svitramia Cham. 2
Tibouchina Aubl. 12
Tibouchineae Cogn. 1, 17
Tibouchinoideae 12
Tristemma Juss. 12